Instructor's Resource Manual with Test Bank to Accompany
Drugs and Society
Fourth Edition

Glen Hanson, Ph.D., D.D.S.
University of Utah
Salt Lake City, Utah

Peter J. Venturelli, Ph.D.
Valparaiso University
Valparaiso, Indiana

Jones and Bartlett Publishers
Boston   London

Editorial, Sales, and Customer Service Offices

Jones and Bartlett Publishers
One Exeter Plaza
Boston, MA 02116
617-859-3900
800-832-0034

Jones and Bartlett Publishers International
7 Melrose Terrace
London W6 7RL
England

Copyright © 1995 by Jones and Bartlett Publishers, Inc.

All rights reserved. Instructors using *Drugs and Society, Fourth Edition* by Glen Hanson and Peter J. Venturelli may reproduce these materials for instruction purposes. Otherwise, no part of the material protected by this copyright notice may be reproduced or utilized in any form, electronic or mechanical, including photocopying, recording, or by any information storage and retrieval system, without written permission from the copyright owner.

ISBN 0-86720-915-1

Printed in the United States of America
99 98 97 96 95   10 9 8 7 6 5 4 3 2

Contents

Preface   v

Chapter 1	An Introduction to Drug Use	1
Chapter 2	Explaining Drug Use and Abuse	9
Chapter 3	Drugs, Regulations, and the Law	19
Chapter 4	How and Why Drugs Work	33
Chapter 5	Homeostatic Systems and Drug	45
Chapter 6	CNS Depressants: Sedative-Hypnotics	55
Chapter 7	Alcohol: Pharmacological Effects	65
Chapter 8	Alcohol: A Behavioral Perspective	73
Chapter 9	Narcotics: (Opioids)	83
Chapter 10	Stimulants	93
Chapter 11	Tobacco	105
Chapter 12	Hallucinogens	113
Chapter 13	Marijuana	123
Chapter 14	Drugs and Therapy	131
Chapter 15	Drug Abuse Among Special Populations	141
Chapter 16	Drug Education, Prevention, and Treatment	151

Preface

This Instructor's Resource Manual with Test Bank is written to accompany *Drugs and Society, Fourth Edition*. Each of the sixteen chapters comprising this manual contains the following:

1. **CHAPTER OVERVIEW** The main points of the chapter are summarized so that the instructor can quickly grasp and review what is covered in the text.
2. **TRUE OR FALSE** This section provides true/false questions that survey each chapter.
3. **MULTIPLE CHOICE** These carefully formulated questions vary from content-oriented to more analytical-type questions.
4. **ESSAY QUESTIONS** The essay questions include a well balanced mixture of questions that challenge the students' understanding of the chapters.
5. **SUPPLEMENTARY MEDIA** The last section lists and describes several films and videos related to each chapter.

I would enjoy hearing from you. Any comments, suggestions or critiques you may have regarding this manual will be greatly appreciated and acknowledged in the next edition. Please direct any comments or critiques to: Peter J. Venturelli, Department of Sociology, Heidbrink Hall, Valparaiso University, Valparaiso, IN 46383; phone (219) 464-5306 or facsimile number (219) 464-6851.

For their continual assistance and unfaltering conscientious efforts in helping to produce this fourth edition resource manual with test bank, we are indebted to the following Valparaiso University students: Jeff Heinze and Korie Anderson, (associate project managers and quality control managers); Phil Farsalas, Mark Sibray, Tanya Senne, Dave Saviola, and Dave Castellanos (typists and first-draft question developers); and Nancy Kostur (associate editor). The combined efforts of all these students should prove that the test questions are both relevant and challenging; their help has contributed to the student orientation inherent in this manual.

− Chapter 1 −
An Introduction to Drug Use

The varied expertise of its authors strengthens this fourth edition. The authors specialize in pharmacology and toxicology, social psychology, and sociology, with social work and health perspectives. There are four major topics in this introductory chapter. The first part reviews the dimensions of drug use and drug abuse and discusses the most commonly abused illicit drugs. The second part focuses on the current extent and frequency of use, statistics and trends, and how the mass media influences our perception in the use of drugs.

The third part delves into how people are attracted to drug use. Particular discussions center on: when use leads to abuse, who is likely to become a user, and how widespread drug abuse is. The fourth and final part uses a newer holistic approach to drug use and emphasizes the theme of this new edition, which is that promoting drug use knowledge strengthens our own awareness of drug use and increases our ability to help others who may have problems with drugs.

TRUE OR FALSE

1. A drug is any substance that modifies the nervous system, but not states of consciousness.
   Ans. F p. 4

2. The use of licit drugs, such as alcohol and tobacco have caused far more deaths, violent crimes and other social problems than the use of illicit drugs.
   Ans. T p. 7

3. Drugs that result from recombined chemical structures of already illicit drugs are called recombined solvents.
   Ans. F p. 7

4. Licit refers to drugs such as marijuana, cocaine and heroin.
   Ans. F p. 7

5. The popular use of legal drugs, particularly alcohol and tobacco, has caused far more deaths, illnesses and social problems than the use of illegal drugs.
   Ans. T p. 7

6. Drug misuse refers to the willful misuse of either legal or illegal drugs for recreation or convenience, while drug abuse refers to the inappropriate or unintentional use of prescribed drugs.
   Ans. F p. 8

7. Attempts to regulate drug use were first made around 1942.
   Ans. F p. 9

8. Drug use and abuse have <u>always</u> been part of human society.   Ans. T p. 9

9. Nearly every culture has laws controlling the wide use of drugs.   Ans. T p. 9

10. The first attempts to regulate drug use were made around 1929 B.C.   Ans. F p. 9

11. Nearly every culture has laws controlling the use of a wide range of drugs.   Ans. T p. 9

12. Approximately 60% of the American population over age 12 tried at least one illegal substance during their lifetimes.   Ans. F p. 11

13. Alcohol addiction is the drug problem that affects the greatest number of people.   Ans. T p. 15

14. Most drug abusers don't perceive any psychological advantage when using these drug compounds.   Ans. F p. 19

15. Floaters are those who drift between experimental and compulsive use.   Ans. T p. 19

16. Substance abuse is confined to only certain socioeconomic and occupational groups.   Ans. F p. 20

17. Drug use is more acute and widespread today than in any previous age.   Ans. T p. 20

18. Major substance abuse is confined to only certain socio-economic groups.   Ans. F p. 20

19. Most people know of at least one close friend or family member who abuses drugs.   Ans. T p. 20

20. Some people are immune to the effects of drug use.   Ans. F p. 21

21. The current declines in drug use reflect our current emphasis on viewing health as interrelated with wellness.   Ans. T p. 22

22. Drug use and abuse have been declining since the mid 1970's.   Ans. T p. 22

23. Holistic health requires self-awareness about the use and abuse of drugs.   Ans. T p. 23

24. Drug use has consistently been a part of human society.   Ans. T p. 28

25. Most drug users are unemployed.   Ans. F
    p. 28

26. Women more so than men tend to use alcohol to cope with problems.   Ans. F
    p. 29

27. Physicians cannot decide among themselves what constitutes legitimate use of a drug.   Ans. T
    p. 29

MULTIPLE CHOICE

1. People in the _____ year old age group are by far the heaviest users of both licit and illicit drugs.   Ans. b   p. 3
    a. 14-18
    b. 18-25
    c. 26-35
    d. over 36

2. According to the National Household Survey of Drug Abuse, which age group is the heaviest drug users and experimenters?   Ans. b   p. 3
    a. ages 12-17
    b. ages 18-25
    c. ages 26-35
    d. ages 36 and over

3. Any substance that modifies biological, psychological, or social behavior is called a (an) ____.   Ans. a   p. 5
    a. drug
    b. medicine
    c. over-the-counter drug
    d. gateway drug

4. Which of the following types of drugs are not commonly abused drugs?   Ans. d   p. 7
    a. cannabis
    b. organic solvents
    c. stimulants
    d. none of the above

5. An example of drug misuse would be _____.   Ans. b   p. 8
    a. psychoactive drugs
    b. prescription drugs
    c. over-the-counter drugs
    d. gateway drugs

6. Drugs that lead to the use and abuse of more powerfully addictive drugs are called _____.   Ans. d   p. 9
    a. psychoactive drugs
    b. prescription drugs
    c. over-the-counter drugs
    d. gateway drugs

7. An example of a "gateway" drug would be _____. Ans. c
   a. LSD   p. 9
   b. opium
   c. marijuana
   d. codeine

8. Licit but never illicit-type drugs that lead   Ans. d
   to use and abuse of more powerfully addictive   p. 9
   drugs are called _____.
   a. prescription drugs
   b. gateway drugs
   c. psychoactive drugs
   d. none of the above

9. Which of the following drugs is not a designer Ans. a
   drug?   p. 9
   a. LSD
   b. PCP
   c. Demoral
   d. Ecstasy

10. The first attempt to regulate drug use was made Ans. a
    in _____.   p. 9
    a. 2240 B.C.
    b. 1030 B.C.
    c. 1840 A.D.
    d. 1920 A.D.

11. Which is not a street term for administering an Ans. d
    opiate?   p. 10
    a. Jolt
    b. Skin poppers
    c. Slam
    d. Blow

12. The four classes of legal chemicals are _____. Ans. c
    a. Social drugs, prescription drugs, over-the-   p. 11
       counter drugs and illicit drugs
    b. Social drugs, over-the-counter drugs and
       miscellaneous drugs
    c. social drugs, prescription drugs, over-the-
       counter drugs and miscellaneous drugs
    d. prescription drugs, over-the-counter drugs,
       miscellaneous drugs and illicit drugs

13. In 1989, how many doses of caffeine were consumed Ans. c
    in the United States?   p. 11
    a. 295 million
    b. 500 million
    c. 295 billion
    d. 500 billion

14. The average household owns about _____ drugs. Ans. c
    a. 12 p. 11
    b. 20
    c. 35
    d. 45

15. ____ of the American population over the age of Ans. b
    12 has tried an illegal substance. p. 11
    a. One fourth
    b. One third
    c. One half
    d. Two thirds

16. Alcohol is or has been used by _____ of all Ans. d
    Americans. p. 15
    a. 36%
    b. 50%
    c. 72%
    d. 85%

17. Of the choices below, the most commonly abused Ans. a
    illicit drug is _____. p. 15
    a. Marijuana
    b. Cocaine
    c. Heroin
    d. Alcohol

18. Advertisers invest huge amounts of money in what Ans. b
    type of medium? p. 16
    a. Magazine advertisements
    b. Television commercials
    c. Newspaper advertisements
    d. Billboards

19. People continue to take drugs because _____. Ans. e
    a. people are searching for pleasure p. 17
    b. peer pressure
    c. it enhances religious experiences
    d. A & B only
    e. All of the above

20. What is the principal factor in determining Ans. b
    drug abuse? p. 17
    a. the amount of drug that is taken
    b. the motive for taking the drug
    c. the frequency of use
    d. drug availability

21. People use drugs for the following reasons: Ans. e
    a. achieve pleasure p. 17
    b. relieve stress
    c. peer pressure
    d. increase concentration
    e. all of the above with the exception of d
    f. all of the above

22. _____ refers to the fact that drug abuse is found among all races, religions, and social levels.
    a. Equal-opportunity affliction
    b. A motivational syndrome
    c. Licit versus illicit
    d. Bydistribution

    Ans. a
    p. 21

23. Since the mid 1970's drug use has _____.
    a. increased slightly
    b. decreased
    c. stayed at the same level
    d. increased dramatically

    Ans. b
    p. 22

24. _____ of drug users are employed in a full or part-time job.
    a. 10%
    b. 30%
    c. 50%
    d. 70%

    Ans. d
    p. 17

ESSAYS

1. What is the difference between drug misuse and drug abuse?

2. List and briefly discuss at least four characteristics common among drug users?

3. What does equal opportunity affliction mean? Why is this important for understanding the extent of drug abuse?

4. In your own words, what explanations do the authors give for the link between drug use and mass media advertising?

5. What similarities and differences do you think you would find between drug abusers and drug abstainers?

SUPPLEMENTARY MEDIA

THE ALLURE OF DRUGS. Examines the historical aspects of drug use in various societies and portrays the attitudes and social images of the users. Quotes users of such stimulants as coffee, tea, opium, tobacco, peyote, hallucinogenic mushrooms, and other substances which have been used as sensory extenders from 4000 B.C. to the 1970s. Indiana University Audio-Visual Center.

AMERICA HURTS: THE DRUG EPIDEMIC. Explores the problem of drug use in America, examining the alarming increase in the number of people using drugs. Includes a segment on crack. Features Collin Siedo as host. Indiana University Audio-Visual Center.

DRUGS: HELPFUL AND HARMFUL. Defines the term "drug" and compares over-the-counter drugs, prescription drugs and illegal drugs. Describes effects of four types of drugs: stimulants, depressants, narcotics, and hallucinogens. Concludes by addressing psychological factors of drug abuse. Uses humorous motif with invisible lecturer. Indiana University Audio-Visual Center.

DRUG-TAKING AND THE ARTS. This program explores how drugs have influenced artistic production in the course of the last 200 years, focusing on major European and American literary figures and visual artists. Actors portray some of the authors, others speak for themselves, and literary critics, psychiatrists, and substance abuse specialists analyze the effects of drugs on the artists. The programs seek to determine whether drugs help the artist produce better art, and if so, how; whether they give the artist insight or only the illusion of insight; whether different drugs produce different sorts of art; and how certain drugs create specific physiological effects in the brain. (2 parts, 56 minutes each, color) Films for the Humanities and Sciences, Princeton, NJ.

TIMOTHY LEARY: A PORTRAIT IN THE FIRST PERSON. Socrates is his role model, convicted of corrupting youth by questions authority. Timothy Leary is perhaps best known for his efforts to "liberate" Americans from the rigors of tradition and conformity by means of mind- and mood-altering drugs. The drugs earned him a 10-year prison sentence. As this program makes clear, this did not dampen his insouciance at being cheerleader of the forces of anti-conformism, voicer of outrageous sentiments that many share but are too polite to utter, above all champion of the individual and individual thought and expression. (24 minutes, color) Films for the Humanities and Sciences, Princeton, NJ.

– Chapter 2 –
Explaining Drug Use and Abuse

This chapter focuses on explaining why people use drugs from biological, psychological, and sociological theoretical perspectives. Time proven and current research findings are included in support of each perspective.

Biological explanations use genetic theories or the disease model for explaining drug addiction. (The disease model assumes that the motivation to use drugs is akin to an uncontrollable illness.) All of the major biological explanations focus on how drug substances alter the chemistry of the brain. And, how drugs interfere with the functioning of neurotransmitters, which are the chemical messengers used for communication between brain regions. Outgrowths of biological explanations are explained under subheadings, drugs of abuse as positive reinforcers, drugs of abuse and psychiatric disorders, and genetic explanations.

Psychological explanations primarily focus on internal mental states affected by individual predispositions and social and environmental influences. Beginning with some interesting reference to Sigmund Freud and his early attempt to explain drug dependence, this section outlines how the widely accepted <u>DSM-IV or Diagnostic and Statistical Manual</u>, 4th edition (American Psychiatric Association, 1994) defines substance dependence, substance abuse, substance intoxications, and substance withdrawal. This section concludes by discussing: personality and drug use and the role of introversion and extroversion; theories based on learning; conditioning; habituation; Bejerot's addiction to pleasure theory; and sociologically based social psychology learning theories that stress how particular forms of reinforcement can explain drug use and/or drug addiction.

The next section of this chapter explains drug use from a sociological perspective. Saying that sociology views the motivation for drug use is determined by the types and quality of social bonds with significant others in familiar physical surroundings, two major explanations are distinguished. The first major explanation is social influence theory, which is encompassed by social learning, the role of significant others in socialization, labeling, and subculture theories. These theories stress how certain primary peer group formations - groups that share a high amount of intimacy, spontaneity and emotional bonding directly influence and determine the motivation to use mainly illicit drugs.

The second major explanation in this section includes theories that are known as structural influence theories. Focusing on how the larger organization of a society, group or subculture is largely responsible for the motivation to use drugs includes social disorganization and social strain, and control theories. Social disorganization points to the extent of integration in our society and how perceived disorganization diminishes the resistance needed to refuse drug use. On a similar, yet different level of explanation is social strain theory, which believes that the inability to achieve desired goals structured by society results in the motivation to use drugs. Finally the last structural influence theory is control theory. This theory emphasizes that if human beings experience a lack of control (social guidance), the motivation to deviate flourishes. In other words, if left without social controls, the natural tendency is to become involved in such deviant behavior as drug use.

The last section of this chapter attempts to predict the risk factors responsible for drug use. Danger signals of drug abuse and a means for explaining drugs use is delineated through using the well known Kumpfer and Turner article, "The Social Ecology Model of Adolescent Substance Abuse: Implications for Prevention."

TRUE OR FALSE

1. Sigmund Freud believed that the addiction to drugs was an outgrowth of habitual masturbatory activity.
Ans. T
p. 33

2. Drug use among working class groups is not as serious as among professional groups.
Ans. F
p. 34

3. Early drug use is <u>not</u> as likely to lead to long-term use as compared to people who begin using drugs later in life.
Ans. F
p. 34

4. *Sinsemilla* is a less potent type of marijuana meaning "with seeds."
Ans. F
p. 35

5. Eighty-eight percent of the U.S. population use drugs daily in some form.
Ans. T
p. 35

6. Biochemical messengers that cause the impulse from one neuron to be transferred to the next are called neurotransmitters.
Ans. T
p. 36

7. The dopamine system is thought to mediate when a drug overdose occurs.
Ans. F
p. 36

8. Some people find the effects of drug abuse very unpleasant.
Ans. T
p. 36

9. The dopamine system is believed to mediate when a drug overdose occurs.
Ans. F
p. 36

10. Psychiatric disorders are potential risk factors drug addiction.
Ans. T
p. 37

11. Studies clearly indicate that genetic influences affect drug vulnerability by 48%.
Ans. F
p. 37

12. Psychological theories mostly deal with social and environmental causal factors.
Ans. F
p. 38

13. DSM-IV, stands for <u>Drug and Substance Manual, 4th Edition</u>.
Ans. F
p. 39

14. The process of getting used to certain patterns of behavior is known as conditioning.
Ans. T
p. 40

15. "Addiction to Pleasure" Theory assumes that it is *psychologically* normal to continue a pleasure stimulus once it is begun.
Ans. F
p. 40-41

16. Sociology views the motivation for drug use as largely determined by the types are quality of bonds with significant others.
Ans. T
p.42

17. Social Learning Theory explains drug use as a result of inward motivational drives.
Ans.F
p. 42

18. Amotivational syndrome that results from drug use causes a lack of ambition and interest in pursuing goals
Ans. T
p. 44

19. Labeling Theory clearly explains why initial drug use occurs.
Ans. F
p. 46

20. The labeling theory suggests that we have little control over the image we have of ourselves.
Ans. T
p. 46

21. According to social learning theory, groups that share a high amount of intimacy, spontaneity, and emotional bonding are called secondary groups.
Ans. F
p. 48

22. Repeated illegal behavior is an example of primary deviance rather than secondary deviance.
Ans. F
p. 48-49

23. Retrospective Interpretation is the social psychologicalprocess of defining a person's reputation based on his/her family upbringing.
Ans.F
p.49

24. Being identified as a doctor, a lawyer, an alcoholic, and HIV positive, are examples of a person's master status.
Ans. T
p. 49

25. Redefining a person's image within a particular group is referred to as retrospective interpretation.
Ans. T
p. 49

26. Social structure theories assert that people do not simply behave on their own but that social behavior results from how subcultures, groups, and society are organized.
Ans. T
p. 50

27. Alienation from society has a negligible effect on the likelihood of deviant behavior.
Ans. F
p. 50

28. Strain theories focus on how the inability to accomplish desired goals leads to drug abuse.
Ans. T
p. 50-51

29. According to social disorganization and strain theories the reason people use and abuse drugs is largely unpredictable.
Ans. F
p. 50-51

30. Loneliness and lack of family support are contributing factors for drug misuse among the elderly.
Ans. T
p. 51

31. Rapid social change does not by itself result in widespread drug use.
Ans. T
p. 51

32. Behavior that is largely dictated by custom and tradition is called sequential behavior.
Ans. F
p. 51

33. Control theories emphasize how drug use in certain groups is normal socialization.
Ans. F
p. 52-53

34. Socialization is the learning process responsible for becoming human.
Ans. T
p. 52

35. Control Theory depicts how conformity to conventional norms and groups prevents deviance.
Ans. T
p. 52-53

MULTIPLE CHOICE

1. Most drug use is _____.
   a. learned by oneself
   b. inherit in genetic makeup
   c. learned from others
   d. learned from parents
Ans. c
p. 33

2. _____ theory can best explain why most use drugs.
   a. genetic
   b. biophysical
   c. psychological
   d. none of the above
Ans. d
p. 33

3. Today, drug use is found _____
   a. in nearly every social group
   b. only in isolated populations
   c. mostly in the inner city
   d. mostly in the suburbs
Ans. a
p. 34

4. Which of the following is <u>not</u> true regarding drug use and abuse?
   a. it increases the possibility of serious accidents
   b. it is commonplace in today's society
   c. it affects nearly every social group
   d. it has steadily declined since the 1960s
   e. none of the above

   Ans. d
   p. 34

5. For which reason(s) do people use drugs?
   a. it makes them feel good
   b. relieves stress
   c. peer pressure
   d. to enhance a religious experience
   e. all of the above

   Ans. e
   p. 35

6. The central nervous system (CNS) is composed of _____ and _____.
   a. the spinal cord; inhibitory system
   b. the spinal cord; the brain
   c. the brain; pituitary gland
   d. the spinal cord; hypothalamus

   Ans. b
   p. 36

7. _____ is/are the brain transmitter believed to medite the rewarding aspects of most drugs of abuse.
   a. neurotransmitters
   b. dopamine
   c. endomophine
   d. synapsis

   Ans. b
   p. 36

8. Sigmund Freud felt that drugs fulfilled insecurities that stem from _____.
   a. chronic masturbation
   b. narcissism
   c. lack of social bonding
   d. parental inadequacies

   Ans. d
   p. 38

9. The DSM-IV discusses mental disorders that result from the use and abuse of all but which drug?
   a. Narcotics
   b. PCP
   c. Alcohol
   d. Caffeine
   e. The DSM-IV discuss all of the above

   Ans. e
   p. 39

10. Which is not a major distinguishing feature of substance abuse. Ans. c
    a. Dependence p. 39
    b. Withdrawal
    c. Abuse
    d. Intoxication
    e. All of the above

11. _____theory assumes that it is biologically normal to continue an enjoyable stimulus. Ans. d
    p. 40

    a. Control
    b. Competency
    c. Subculture
    d. Addiction to pleasure

12. Social learning theory emphasizes that _____. Ans. b
    a. drug taking behavior is omnipresent p. 41
    b. learning occurs through intimate interaction with others
    c. learning occurs through unconscious motivation
    d. drug taking behavior is partly learned, partly genetic, and also occurs because of unknown

13. Which is not considered a social influence theory? Ans. a
    a. social conflict p. 42
    b. labeling
    c. subculture
    d. role of significant others

14. Macroscopic explanations are _____. Ans. d
    a. closeup explanations p. 42
    b. detailed explanations
    c. day to day explanations for drug use
    d. comprehensive explanations

15. According to the social learning theory which is an example of a primary group? Ans. c
    a. clerks at grocery stores p. 42
    b. acquaintances
    c. step-parents
    d. employees at department stores who know you by sight but not by name

16. Which of the following, according to Becker, occcurs first with regard to learning drug use? Ans. a
    p. 43

    a. learning the technique
    b. learning the preuse effects
    c. perceiving the effects
    d. learning where and from whom the drug can be purchased

17. _____ is the drug that has been used by the highest percentage of high school students
    a. Alcohol
    b. "Crack"
    c. Cigarettes
    d. Marijuana

    Ans. a
    p. 44

18. After drug use has begun continuing behavior does not involve _____.
    a. who to purchase the drug from
    b. where to get the money to purchase
    c. how to maintain the secrecy of use
    d. the justification for continual use

    Ans. b
    p. 44

19. According to Coleman, the adolescent period emphasizes all of the following except _____.
    a. a tendency toward unconventional behavior
    b. psychic attachment to one another
    c. a tendency toward conventional behavior
    d. a drive toward autonomy

    Ans. c
    p. 45

20. The pressure to conform is more strongly felt and perceived as a threat by the _____.
    a. old
    b. young
    c. middle-aged
    d. nonconformers

    Ans. b
    p. 45

21. According to _____ theory, if someone is perceived as a drug user, this perception functions as a label of that person's character and affects his or her self-perception.
    a. social learning
    b. labeling
    c. subculture
    d. social strain

    Ans. b
    p. 46

22. According to _____ theory, if someone is perceived as a drug user, this perception functions as a label of that person's character and affects his or her self perception.
    a. structural
    b. subculture
    c. social strain
    d. labeling

    Ans. d
    p. 46

23. Drug use is generally more prevalent in adolescents who _____.   Ans. d   p. 47
    a. maintain close family ties
    b. question the values of conventional society
    c. hold the values and general attitudes of parents
    d. have poor family relationships

24. According to Hall, et al., physicians have one of the highest addiction rates (to drugs other than alcohol) because _____.   Ans. c   p. 48
    a. they know about the medical affects of drugs
    b. drugs help with the personal family problems of doctors
    c. they have easy access to drugs
    d. none of the above

25. One underlying factor that distinguishes secondary from primary deviance is _____.   Ans. d   p. 48
    a. enjoying the deviance
    b. becoming attached to rule-breaking behavior
    c. feeling trapped
    d. identification with the deviance

26. The International Olympic Committee has prohibited the use of all of the following except _____.   Ans. d   p. 49
    a. anabolic steroids
    b. depressants
    c. stimulants
    d. ethanol

27. Secondary deviance is or primarily involves _____.   Ans. d   p. 49
    a. inconsequential deviant behavior
    b. when the perpetrator does not identify with the deviance
    c. usually first-time violators of law
    d. when the perpetrator identifies with the deviant behavior

28. _____ theories explain drug use as being caused by group influence and peer pressure.   Ans. b   p. 49
    a. Social disorganization
    b. Subculture
    c. Social control
    d. None of the above

29. _____ theories explain drug use as being caused by group influence and peer pressure.   Ans. a   p. 49
    a. subculture
    b. social disorganization
    c. social control
    d. none of the above

30. The most significant predictor of marijuana and cocaine use among employees is _____.
    a. age
    b. occupation
    c. sex
    d. family background

    Ans. a
    p. 50

31. A similarity between elderly and adolescent drug abusers is that they _____.
    a. lack income
    b. lack self-reliance
    c. possess limited amount of social status
    d. all of the above

    Ans. d
    p. 51

32. _____ describes a situation where, because of rapid social change, previously affiliated individuals no longer fine themselves integrated into acommunities social institutions.
    a. social strain theories
    b. social disorganization theories
    c. competency theories
    d. economic and social liability theories

    Ans. b
    p. 51

33. Which of the following is not a reason the elderly begin to abuse drugs?
    a. aging process
    b. peer influence
    c. alienation
    d. illness

    Ans. b
    p. 51

34. _____ theories identify how rapid social change is socially disruptive.
    a. Labelling
    b. Social disorganization and strain
    c. Frustration-aggression
    d. Control

    Ans. b
    p. 53

35. Travis Hirschi believes that delinquent behavior (which includes drug use) has a tendency to occur whenever people lack _____.
    a. attachment to others and commitment to goals
    b. involvement in conventional activity
    c. belief in a common value system
    d. all of the above

    Ans. d
    p. 53

36. _____ theory explains the socialization process as resulting from the creation of strong or weak internal and external control systems.
    a. frustration-aggression
    b. conflict
    c. containment
    d. social control

    Ans. c
    p. 53

ESSAYS

1. How do social structure theoretical explanations differ from social process theories? Give an example of each theory for explaining drug abuse.

2. Contrast and compare subculture vs. labeling theories. How does each theory explain drug use? How do they differ in their explanations?

3. List and describe three factors in the learning process that Howard Becker believes first-time users go through before they become attached to using illicit psychoactive drugs.

4. Explain the link between rapid social change, the proliferation of subcultures and increased drug use?

5. Explain how internal vs. external control works and give an example of each.

6. In your own words, explain the following concepts and give a drug related example of each: primary and secondary deviance, master status, and retrospective interpretation.

SUPPLEMENTARY MEDIA

WHATEVER HAPPENED TO CHILDHOOD. This forty-six minute videotape shows that childhood, as today's adults knew it, no longer exists. Early exposure to the pressures of drug usage and changing sexual morality, along with a new indifference to and participation in criminal behavior is forcing our children to become adults at a younger age. Both young people and adults tell their stories and explore our changing society in this powerful fast-moving documentary. Churchill Films. Available from Kent State Film Library, Kent, Ohio 44242.

DUAL DIAGNOSIS (1993). This 20 minute video explores why people, who suffer from mental disorders as well as drug addiction, frequently relapse to drug use following a short period of abstinence. National Clearinghouse for Alcohol and Drug Information (NCADI), P.O. Box 2345, Rockville, MD 20847-2345 or telephone (800) 729-6686.

THEORY: CAUSES OF DRUG USE. Drug trade and drug use is a problem of global proportions. Should society treat drug abuse as a health problem, or a criminal justice problem? This two-part program shows how the drug problem is being addressed in four major cities: New York, Toronto, Amsterdam and Liverpool. In Amsterdam and Liverpool drug users are treated as people who are sick, not evil. They are given needles, syringes, medical care and in some cases the drug itself. Street crime has dropped. Interviews with addicts show that when they have a steady supply of the drug they are able to lead a more normal life and not resort to crime and prostitution. New York is a city in chaos, racked by drugs and the effort to root them out. Stern drug laws have not provided an answer. A prosecutor complains that the war on drugs has diverted police and court resources away from crimes like murder and rape. Also, incarcerating small-time drug users in overcrowded prisons does not cure their addiction. Canada's public policy is somewhere between Europe's liberal approach and America's hard line. While drug use has declined in the general population, there has been a rampant spread among street kids. This film questions our underlying assumptions about illegal drug use and offers a new approach to this devastating social ill. (94 minute part I and II). Filmakers Library New York, NY 10016.

− Chapter 3 −
Drugs, Regulations, and the Law

The chapter describes how drug regulation has evolved in the United States and the role of substances of abuse in that evolutionary process. A hundred years ago our government was more concerned about protecting the rights of drug manufacturers than providing for public safety. As the dangers and fraudulent claims of the early, uncontrolled "patent" medicines became apparent, the government began to pass legislation to regulate drug safety and effectiveness. The Pure Food and Drug Act of 1906, although ineffective, was the beginning of involvement by governmental agencies in drug manufacturing. This Act made it illegal to misrepresent intentionally a drug product on its label. The Act was in response to growing problems of drug addiction and its underlying purpose was to keep narcotic-or-cocaine-containing patent medicines from being advertised as nonhabit forming.

From subsequent legislation (1) the Food and Drug Administration (FDA) was created as the principal drug regulatory agency for the federal government, (2) drugs were classified into prescription and nonprescription types, (3) approved drugs were required to be both relatively safe and effective, and (4) new drugs were required to be thoroughly tested before marketing was allowed.

Special attention is given to the regulation of substances of abuse in this chapter. Those drugs that have abuse potential are regulated in a manner distinct from other drugs; chapter three explains why and what these differences are. The ways in which law enforcement agencies deal with substance abuse are largely determined by the Comprehensive Drug Abuse Prevention and Control Act of 1970. The chapter discusses how drugs with abuse potential are classified into Schedules ranging from I to V, according to their likelihood of being abused and their clinical usefulness and how regulation has affected the patterns of drug abuse, the drug abuser and society. In addition, this chapter includes discussions of (1) the linkage between crime and substance abuse, (2) the controversy of legalizing the use of substances of abuse, and (3) the impact of drug testing on drug abuse issues.

TRUE OR FALSE

1. The U.S. government has always regulated product quality and claims, in regards to drugs.
   Ans. F
   p. 62

2. Patent medicines at the turn of the century were not effectively regulated by the government.
   Ans. T
   p. 63

3. Most doctors in the 1800s were not aware of the dangerous properties of opium.
   Ans. F
   p. 64

4. George Wood, a prestigious University of Pennsylvania professor, announced that opium dulled the intellectual and imaginative faculties of a person.
Ans. F
p. 65

5. In the late 1800s opium and cocaine were used as "cures" for addiction to other drugs.
Ans. T
p. 66

6. During the 1800s, cocaine was often used to wean opium addicts from their narcotic habits.
Ans. T
p. 66

7. Unlimited quantities of brandy and champagne early suggested by Edward Levinstein as treatment for withdrawal symptoms in patients attempting to get rid of an opium addition.
Ans. T
p. 66

8. The lack of safety and effectiveness of the early patent medicines significantly contributed to the passage of the 1906 Pure Food and Drug Act.
Ans. T
p. 67

9. Coca-Cola was once a tonic containing addictive cocaine.
Ans. T
p. 67

10. The Sherley Amendment of 1912 allowed the manufacturer to determine whether a drug was to be labeled prescription or nonprescription.
Ans. F
p. 67

11. The Kefauver-Harris Amendment of 1962 required drug companies to prove their new medications were safe but not necessarily effective before they could be marketed.
Ans. F
p. 69

12. The Drug Efficacy Study started in 1966 rated the effectiveness of drugs.
Ans. T
p. 70

13. The "initial clinical stage" or phase 1 testing is conducted in laboratory animals.
Ans. F
p. 71

14. Phase 3 testing of an Investigational New Drug (IND) is referred to as the "clinical pharmacological evaluation stage."
Ans. F
p. 71

15. Most new drugs require 2-4 years to be tested and approved for marketing.
Ans. F
p. 72

16. The "Orphan Drug Law" gives tax advantages to pharmaceutical companies producing drugs to treat high blood pressure.
Ans. F
p. 72

17. When the FDA began evaluating OTC drugs in 1972 for safety and effectiveness, it found that only approximately 700 different active ingredients were included in more than 300,000 drug products.
Ans. T
p. 73

18. Once a drug is designated as prescription, the FDA is not allowed to switch it to OTC status at a later time.
Ans. F
p. 73

19. The FDA's switching policy refers to allowing OTC products to be reviewed and possibly be changed to prescription status.
Ans. F
p. 73

20. The extent to which the advertisement fails to reveal facts is considered in the determination of false advertisement.
Ans. T
p. 74

21. The Food and Drug Administration (FDA) is the federal agency primarily responsible for regulating the advertisement of prescription drugs.
Ans. T
p. 74

22. The Harrison Act of 1914 was one of the first legitimate efforts by the government to regulate and control substances of abuse.
Ans. T
p. 75

23. Drug laws are usually more lenient to the seller than to the user of the drug of abuse.
Ans. F
p. 77

24. The availability of drugs has diminished substantially due to interdiction.
Ans. F
p. 78

25. Illicit drugs that are chemically modified so they are not considered illegal but still retain abusive properties are called "designer" drugs.
Ans. T
p. 80

26. As of 1989 approximately 20% of the AIDS patients were intravenous drug users.
Ans. T
p. 81

27. It is not likely that use of amphetamines or cocaine during pregnancy can cause damage to the fetus.
Ans. F
p. 81

28. In some drug-related criminal proceedings, the insanity defense is used.
Ans. T
p. 83

29. Drug abuse was seen as the number one problem in the U.S. in 1989 and 1990 by the majority of its citizens.
Ans. T
p. 83

30. The so-called "war on drugs" has been proven to be the principal reason for the decline in drug abuse in the early 1990s.
Ans. F
p. 84

31. The only people advocating legalization of drugs of abuse are individuals who are themselves dependent on these substances.
Ans. F
p. 84

32. It has been proven that legalization of marijuana will reduce its use and social problems.
Ans. F
p. 84

33. Most people in the U.S. want to legalize most substances of abuse.
Ans. F
p. 85

34. Restrictive laws are relatively ineffective in reducing drug abuse and addiction.
Ans. T
p. 88

35. Drug testing as currently used has been proven to be an effective deterrent to drug abuse in the general population.
Ans. F
p. 88

36. The Alcohol and Drug Abuse Education Amendments placed more emphasis on drug abuse in rural areas.
Ans. T
p. 92

37. A major reason that President Nixon established the Special Action Office For Drug Abuse Prevention (SAODAP) was the report of a high heroin addiction rate in returning Vietnam veterans.
Ans. T
p. 93

38. State law enforcement of drug statutes is consistent from state to state.
Ans. F
p. 94

39. All new drugs to be considered for marketing must be first tested on two species of animals.
Ans. F
p. 96

MULTIPLE CHOICE

1. Collier's Magazine coined the phrase "dope fiend" which comes from dope, an African word meaning _____.
    a. "illegal substance"
    b. "stupid person"
    c. "intoxicating substance"
    d. "drug user"
Ans. c
p. 64

2. In 1905 _____ Magazine ran a series of articles called "Great American Fraud," which dealt with the abuse of patent medicines.
    a. Collier's
    b. New Yorker
    c. Time
    d. Life
Ans. a
p. 64

3. Laudanum, which contained opium, was commonly used in the 1800s to _____.
    a. treat pain
    b. promote a sense of relaxation
    c. treat diarrhea
    d. all of the above
Ans. d
p. 64

4. In 1842, a tax was place on the importation of _____.
   a. marijuana
   b. opiates
   c. alcohol
   d. cocaine

   Ans. b
   p. 66

5. Which of the following was initiated by the Harrison Act of 1914?
   a. increased penalties levied against illegal use of opiates
   b. marijuana was viewed as different from other narcotic drugs
   c. dealers of narcotics were required to register annually
   d. the formation of the Food and Drug Administration

   Ans. a
   p. 66

6. The Pure Food and Drug Act of 1906 accomplished all of the following, except:
   a. initiating the decline of patent medicines
   b. marking the beginning of government involvement in drug manufacturing
   c. labeling the amount of addictive drugs in a product
   d. declaring addictive drugs illegal

   Ans. d
   p. 67

7. Select the incorrect statement concerning the "Pure Food and Drug Act of 1906."
   a. The increased awareness of the dangers of opium and cocaine helped lead to passage of this act.
   b. It required that potentially addicting patent medicines not be advertised as "nonhabit forming".
   c. The Sherley Amendment to this act made it illegal to sell ineffective drugs.
   d. The Sherley Amendment did not significantly improve the quality of drug products.

   Ans. c
   p. 67

8. Which of the following did not occur as a direct result of the Federal Food, Drug and Cosmetic Act of 1938?
   a. it created a class of drugs that could be obtained with a prescription
   b. it required all drugs to be proven effective before marketed
   c. it allowed the manufacturer to determine whether a drug was to be a prescription or nonprescription medication
   d. it led to the Durham-Humphrey Amendment

   Ans. b
   p. 67

9. Select the incorrect association. Ans. c
   a. The Pure Food and Drug Act of 1906 was a p. 67
      response to patent medicines containing
      addictive drugs.
   b. The Sherley Amendment of 1912 was a response
      to false claims made from a cancer "remedy."
   c. The Federal Food, Drug, and Cosmetic Act of
      1938 was passed because of the marketing of
      ineffective antibiotics.
   d. The Durham-Humphrey Amendment of 1951 was
      passed because the FDA and drug companies
      thought the new tranquilizers and antibiotics
      of the time were too dangerous to be sold
      without a prescription.

10. The Food, Drug, and Cosmetic Act of 1938... Ans. a
    a. defined drugs to include products that p. 68
       affected body structure or function in the
       absence of disease
    b. declared that companies did not have to file
       applications for all new drugs
    c. stated that the drug label did not have to
       include all ingredients and the quantities
       of each
    d. declared that drug warnings were not required
       to appear on the drug label

11. An outbreak of birth defects caused by the use Ans. b
    of thalidomide during pregnancy led to the p. 69
    passage of _____.
    a. the Food, Drug, and Cosmetic Act
    b. the Kefauver and Harris Amendments
    c. the Durham-Humphrey Amendment
    d. none of the above

12. Which of the following legislation occurred Ans. a
    because of the thalidomide tragedy? p. 69
    a. Kefauver and Harris Amendment
    b. Pure Food and Drug Act
    c. Federal Food, Drug and Cosmetic Act
    d. Harrison Act

13. _____ means "able to cause abnormal development Ans. b
    of the fetus." p. 70
    a. Carcinogenic
    b. Teratogenic
    c. Mutagenic
    d. Paragenic

14. During which phase of human testing does an investigational drug become available on a wide experimental basis?
    a. Phase 1
    b. Phase 2
    c. Phase 3
    d. Phase 4

    Ans. c
    p. 72

15. The Orphan Drug Law was introduced to promote the development and testing of drugs to be used in the treatment of _____.
    a. orphan children
    b. rare diseases
    c. high incident diseases
    d. diseases requiring expensive treatments

    Ans. b
    p. 72

16. Switching prescription drugs to OTC status presents the potential problem of:
    a. raising medical costs
    b. a public ignorant about these agents and their side effects
    c. doctors losing money
    d. none of the above

    Ans. b
    p. 73

17. The Comprehensive Drug Abuse prevention and Control Act of 1970 emphasized the area of _____.
    a. law enforcement
    b. research
    c. education
    d. treatment

    Ans. a
    p. 75

18. From 1987 to 1990 there was a dramatic decline in demand for ____ in the United States.
    a. alcohol
    b. opium
    c. LSD
    d. cocaine

    Ans. d
    p. 79

19. In which of the following states has it been claimed that marijuana is the single largest cash crop?
    a. Florida and California
    b. California and New York
    c. New York and Florida
    d. California and Oregon

    Ans. d
    p. 79

20. What are the immeasurable costs society pays for drug addiction?
    a. loss of good minds to industry and profession
    b. shortened lives
    c. few illnesses
    d. both a and b
    d. all of the above

    Ans. d
    p. 80

21. Most crimes related to drugs involve _____. Ans. b
    a. homicide p. 80
    b. theft of personal property
    c. assault and robbery
    d. fraud

22. A major cost of drug addiction is _____. Ans. d
    a. treating drug-related emergencies p. 80
    b. paying for the effects of drug-related crimes
    c. rehabilitating users
    d. all of the above

23. As of 1993, what percent of AIDS cases in Ans. b
    the United States were blamed on intravenous (IV) p. 81
    drug users?
    a. 20%
    b. 25%
    c. 45%
    d. 60%

24. The most common defense used in drug-related Ans. a
    criminal proceedings is: p. 82
    a. individual liberty
    b. right to privacy
    c. freedom of religion
    d. freedom of speech

25. Of the illegal drugs for abuse, the one most Ans. c
    likely to be legalized is _____. p. 86
    a. cocaine
    b. "crack"
    c. marijuana
    d. LSD

26. Which of the following was excluded from the Ans. c
    Drug Abuse Control Amendment of 1965? p. 90
    a. LSD
    b. barbiturates
    c. marijuana
    d. amphetamines

27. An Investigational New Drug (IND) for treating Ans. b
    high blood pressure is being given to a few p. 71
    patients with hypertension to check its
    effectiveness. This drug is most likely found in
    which of the following phases of human testing?
    a. Phase 1
    b. Phase 2
    c. Phase 3
    d. Phase 4

28. Which of the following best describes the "fast-track" rule for drug marketing? Ans. b
    a. provides special tax incentives for drugs to be used on rare diseases p. 72
    b. allows rapid evaluation and early marketing for drugs with important and obvious therapeutic action
    c. allows rapid evaluation for drugs to be marketed OTC
    d. slows evaluation for drugs shown to have serious side effects in some populations

29. Select the correct statement concerning the regulation of OTC drugs. Ans. c
    a. The FDA has thoroughly evaluated each of the more than 300,000 OTC drug products available. p. 73
    b. The FDA began evaluating the effectiveness and safety of OTC products in 1951.
    c. The active ingredients in OTC medicines have been classified in one of three categories according to their effectiveness and safety.
    d. Some safe and effective prescription drugs are being reclassified as OTC medications.

30. An OTC ingredient found to be both safe and effective is likely to be classified by the FDA in which of the following categories? Ans. a
    a. Category I p. 72
    b. Category II
    c. Category III
    d. Category IV

31. Which of the following best describes "the switching policy" of the FDA? Ans. a
    a. changing of prescription drugs to OTC status p. 73
    b. changing of OTC drugs to prescription status
    c. a new FDA hiring policy designed to put more minorities and females in administrative positions
    d. a policy trend at FDA to become more rigid in its enforcement of federal drug laws

32. Which federal agency is primarily responsible for regulating the advertising of OTC drugs? Ans. a
    a. Federal Trade Commission (FTC) p. 73
    b. Food and Drug Administration (FDA)
    c. The Better Business Bureau
    d. Drug Enforcement Administration (DEA)

33. Drugs that have high abuse potential and can also be prescribed by health professionals with some restrictions are usually classified in which of the following schedules?
    a. I
    b. II
    c. III
    d. IV

    Ans. b
    p. 75

34. Select the <u>incorrect</u> statement concerning the use of interdiction programs to stop drug abuse.
    a. In the U.S. they usually receive more funding than programs to treat drug addicts.
    b. This strategy has been proven to dramatically decrease demand for substances of abuse.
    c. If it is successful in reducing the supply of a drug, that drug is often replaced by another with similar abuse potential.
    d. In spite of interdiction, if the price is high enough, a way is found to satisfy the drug demand.

    Ans. b
    p. 78

35. Which of the following best describes the general drug abuse pattern in adolescents from 1987 to 1990?
    a. a dramatic decline
    b. a gradual decline
    c. no change
    d. a gradual increase

    Ans. a
    p. 79

36. Select the <u>incorrect</u> statement.
    a. It is likely that between $50 to $100 billion is spent each year for illegal substances of abuse in the U.S.
    b. The largest crop exported from Bolivia and Colombia are illegal substances of abuse.
    c. The typical heroin addict usually spends less than $10 a day on his or her habit.
    d. 20 to 30% of the prostitutes in major cities are dependent on drugs and use prostitution to support their drug habit.

    Ans. c
    p. 80

37. Select the <u>incorrect</u> statement.
    a. Clandestine laboratories throughout the country synthesize illicit amphetamine-like and heroin-like drugs.
    b. The clandestine laboratories usually are careful in their techniques and prepare relatively pure illicit drugs.
    c. Contaminants of illicit drugs synthesized by clandestine laboratories can be toxic and cause death in users.
    d. MPTP is a contaminant of a "synthetic heroin" that can cause Parkinson's disease.

    Ans. b
    p. 80

38. Which of the following illicit drug is most likely to be legalized in some states?
    Ans. b
    p. 86
    a. cocaine
    b. marijuana
    c. LSD
    d. heroin

ESSAYS

1. Describe the two major criteria that determine how society regulates drugs.

2. List the three principal stages of human testing for Investigational New Drugs (IND) and describe the nature and purpose of the testing for each stage.

3. List the criteria that must be met if a drug is to be switched by the FDA from prescription to OTC status.

4. List and discuss three strategies which can be used to reduce the demand for substances of abuse.

5. Describe the major legal defenses used in drug-related crimes.

6. Describe three significant reasons for persistent drug-abuse problems despite stricter law enforcement.

7. Describe the advantages and disadvantages of legalizing the use of cocaine; of heroin; of marijuana.

8. What are the major problems of trying to prevent use of illicit substances with drug testing strategies?

SUPPLEMENTARY MEDIA

DRUG TESTING IN THE WORKPLACE. This twenty-three minute videotape points out the importance of drug testing to ensure that all employees are able to work in a safe and secure workplace. Dispels the myths and misconceptions about drug testing, demonstrates urine and blood testing in the lab, and deals with the issues of confidentiality, prevention of tampering, and rehabilitation. Presents one company's policy for drug screening, showing how it goes into effect after an on-the-job accident, and follows the worker who caused the accident through all the interviewing and testing procedures. Stresses that drug testing is not a law enforcement tool but a diagnostic tool designed to help those who need it and to prevent accidents on the job. Available from Indiana University Audio-visual Center, Bloomington, Indiana 47405-5901.

BUSTED. This short film looks at the threat of arrest and jail for those persons using drugs illegally. Available from Kent state Film library, Kent, Ohio 44242.

DRUGS AND EVIDENCE. This film discusses the legal aspects of drug use and abuse and the identification of drug abuse. Available from the Kent State film Library, Kent, Ohio 44242.

DRUG TESTING: HANDLE WITH CARE. Audience: adults. Length: 22 minutes. This video presents information about the nature of substance abuse in the workplace. The video stresses the importance of the four components of an effective workplace program: education, an employee assistance program, supervisor training, and drug testing. Employers' and employees' versions (VHSO2 and VHSO5). NCADI (P.O.Box 2345, Rockville, MD 30847-2345).

— Chapter 4 —

HOW and WHY DRUGS WORK

This chapter discusses those factors that determine how drugs will affect the user. The significance of drug dose to potency, toxicity, clinical usefulness and abuse potential is described. The types of drug interactions and their implications are presented. The chapter describes pharmacokinetic factors that influence drug responses. These include how drugs are administered; how and where drugs are distributed once they enter the body; how the body inactivates and eliminates drugs from the body. Other concepts discussed that help students understand how drugs work are rebound effects, sensitization, tolerance, physical and psychological dependence or addiction. A discussion of the factors that lead to drug dependencies concludes the chapter.

TRUE OR FALSE

1. Most drugs can cause nausea or an upset stomach in someone. — Ans. T p. 102

2. Psychological dependence on a drug often leads to drug abuse. — Ans. T p. 102

3. Everyone does not respond the same to a given dose of a drug. — Ans. T p. 104

4. The ED50 of a drug refers to the effect of the drug when 50mg are administered. — Ans. F p. 105

5. The LD50 is almost always greater than the LD1. — Ans. T p. 105

6. It would be preferred to use a drug with a therapeutic index of 50 rather than a drug with a therapeutic index of 1. — Ans. T p. 106

7. The lower the therapeutic index, the safer the drug. — Ans. F p. 106

8. The smaller the dose required to achieve a drug action, the greater the drug potency. — Ans. T p. 106

9. Drug effects can be altered by interacting with food. — Ans. T p. 108

10. Many drugs influence the actions of the other drugs. — Ans. T p. 108

11. The analgesic effects of aspirin and Tylenol are addictive when the two drugs are combined. — Ans. T p. 108

12. The smaller the therapeutic index, the safer the drug. — Ans. F p. 108

13. The higher the therapeutic index, the more desirable the drug. — Ans. T p. 108

14. There is no such thing as "the perfect drug". — Ans. T p. 108

15. Alcohol is a CNS depressant. — Ans. T p. 109

16. The CNS depressant effects of alcohol and barbiturates interact in an additive fashion. — Ans. F p. 109

17. Most drug abusers are multiple drug users. — Ans. T p. 110

18. The form in which a drug is administered does not influence the rate of passage into the bloodstream from the site of administration. — Ans. F p. 110

19. The form in which a drug is taken does not influence the effect it has. — Ans. F p. 110

20. Some drugs cannot be administered orally because they are inactivated in the stomach before being absorbed into the blood. — Ans. T p. 111

21. Norplant is a contraceptive administered as an implantable drug delivery system. — Ans. T p. 112

22. It requires about 10 minutes for a drug administered I.V. to circulate completely throughout the body. — Ans. F p. 113

23. Most drugs are distributed throughout the body in the blood. — Ans. T p. 113

24. All drugs can enter the brain once they get into the blood. — Ans. F p. 113

25. Most drugs do not pass from the mother into the fetal circulation. — Ans. F p. 113

26. Most drugs cross the placental barrier from the mother to the fetus. — Ans. T p. 113

27. Immediate or short-term effects after taking a single drug dose is called chronic. — Ans. F p. 114

28. The liver is the major organ that metabolizes drugs and hormones in the body. — Ans. T p. 114

29. The plateau effect is the maximum effect of a drug, regardless of its dose.
Ans. T
p. 114

30. The effects of acute and chronic treatments with the same drugs are always the same.
Ans. F
p. 114

31. The shorter the time interval between doses of a drug the greater the likelihood of a cumulative effect from that drug.
Ans. T
p. 114

32. Biotransformation does not affect the potency of drugs.
Ans. F
p. 114

33. Individuals with a history of cardiovascular disease take a greater risk with OTC decongestants than other users.
Ans. T
p. 116

34. Because of cross-dependency, barbiturates can be used to treat the abstinence syndrome of the chronic alcoholic.
Ans. T
p. 111

35. The mental set can profoundly influence a person's response to some drugs.
Ans. T
p. 121

36. The set and setting are not particularly important factors in determining responses to drugs of abuse.
Ans. F
p. 121

37. Tolerance to one drug can often cause tolerance to other similar drugs; this is called reverse-tolerance.
Ans. F
p. 119

38. Drug tolerance causes an increased response to a given dose of a drug.
Ans. F
p. 117

39. Withdrawal effects usually occur when drug use is stopped in persons who are physically dependent on that drug.
Ans. T
p. 119

40. If tolerance develops to the effects of one drug, often tolerance has also developed to other similar drugs.
Ans. T
p. 119

41. Psychological dependence rarely leads to drug abuse.
Ans. F
p. 120

42. Most people who have used psychoactive drugs do not develop significant psychological dependence.
Ans. T
p. 120-1

43. Offspring of alcoholics are no more likely to develop alcohol dependency than children of non-alcoholic parents.
Ans. F
p. 123

44. Effects caused by suggestion and psychological factors, not the pharmacological activity of a drug, are referred to as placebo effects.
Ans. T
p. 121

45. Pain can often be relieved by a placebo medication.
Ans. T
p. 121

46. The endorphins are chemicals that are produced by the body and mimic the effects of the narcotic drugs.
Ans. T
p. 122

MULTIPLE CHOICE

1. Basic kinds of side effects can include _____.
   a. nausea
   b. changes in mental alertness
   c. dependence
   d. all of the above

   Ans. d
   p. 102

2. Unpleasant effects that occur when use of a drug is stopped is called _____.
   a. potency
   b. withdrawal
   c. metabolism
   d. threshold

   Ans. b
   p. 103

3. Which of the following could contribute to individual variability in drug responses?
   a. rate of metabolism by the liver
   b. acidity of the urine
   c. size of the person
   d. all of the above

   Ans. d
   p. 104

4. Select the incorrect statement.
   a. A very potent drug often reaches toxic levels at low doses.
   b. Potency is determined by the amount of drug necessary to cause a given effect.
   c. A drug is considered to be highly toxic if the LD-50 is between 0.5 and 5 grams/kg.
   d. Toxicity is not always related to dosage in a simple linear relationship.

   Ans. c
   p. 105
   107

5. Which of the following can be toxic at sufficiently high doses?
   a. aspirin
   b. vitamin A
   c. table salt
   d. all of the above

   Ans. d
   p. 106

6. The capacity of a drug to do damage or cause adverse effects in the body is called _____.
   a. toxicity
   b. potency
   c. metabolism
   d. tolerance

   Ans. a
   p. 106

7. The toxic dose (LD-50) divided by the therapeutic dose (ED-50) is used to calculate the_____.
   a. margin of safety
   b. therapeutic index
   c. synergism
   d. pharmacokinetics

   Ans. b
   p. 106

8. _____ is the range in dose between the amount of drug necessary to cause a therapeutic effect on a toxic effect.
   a. Margin of safety
   b. Therapeutic index
   c. Synergism
   d. Pharmacokinetics

   Ans. a
   p. 107

9. Which of the following best describes the concept "margin of safety"?
   a. It compares the advantages and disadvantages of a drug.
   b. It is the same concept as "benefit/risk".
   c. Drugs with steep dose response curves generally have a wide margin of safety.
   d. It is a comparison of dosage for therapeutic and toxic effects.

   Ans. d
   p. 107-108

10. Drug A has a therapeutic index of 100 and drug B has a therapeutic index of 1. What does this mean when comparing these two drugs?
    a. Drug B is a safer drug than A.
    b. Drug A is a more effective drug than B.
    c. Drug B is more effective than A.
    d. Drug A has a wider margin of safety than drug B.

    Ans. d
    p. 108

11. Which of the following interactions best describes the analgesic effect when aspirin is combined with acetaminophen?
    a. potentiation
    b. synergism
    c. antagonism
    d. additive

    Ans. d
    p. 108

12. Pharmacologists refer to a perfect drug as a _____. Ans. c
    a. wonder drug p. 108
    b. super pill
    c. magic bullet
    d. dynamic drug

13. What kind of drug effect occurs when one drug cancels or blocks the effect of another? Ans. b
    a. addictive drug p. 108
    b. antagonistic drug
    c. proactive drug
    d. interceptive drug

14. _____ is the ability of one drug to enhance the effect of another. Ans. d
    a. Haptene p. 109
    b. Dysphoric
    c. Metabolism
    d. Synergism

15. It is estimated that as many as _____ people die each year from mixing alcohol with CNS depressants. Ans. c
    a. 300 p. 109
    b. 500
    c. 3000
    d. 30,000

16. Which of the following best describes synergisic drug interactions? Ans. d
    a. the effects of two drugs sum together p. 109
    b. the effect of one drug does not alter the effect of another drug
    c. the effect of one drug cancels the effect of another
    d. the presence of one drug enhances the effect of another drug

17. In determining how drugs affect the body, pharmacologists need to be concerned with _____? Ans. d
    a. administration p. 110
    b. distribution
    c. biotransformation
    d. all of the above

18. The study of factors that influence the distribution and concentration of drugs in the body is called _____. Ans. a
    a. pharmacokinetics p. 110
    b. pharmacodynamics
    c. teratogenics/macodynamics
    d. none of the above

19. Which of the following statements concerning drugs administered orally is <u>incorrect</u>?
  a. This is usually the most convenient form of drug administration.
  b. Drugs absorbed from the gut do not get to the liver for metabolism until after they are distributed to the general circulation.
  c. Some drugs are inactivated by digestive enzymes in the gut when taken orally.
  d. The onset of action is usually slower for drugs taken orally than for drugs injected I.V.

   Ans. b
   p. 110-112

20. The principal forms of drug administration include which of the following?
  a. injection
  b. oral ingestion
  c. suppository
  d. all of the above

   Ans. d
   p. 110-112

21. Which is usually the fastest way for a drug to affect the body?
  a. intravenous injection
  b. intramuscular injection
  c. orally
  d. suppository

   Ans. a
   p. 112

22. Which of the following forms of administration <u>is the least reliable</u> way of getting most drugs into the blood?
  a. oral
  b. topical
  c. intravenous
  d. intramuscular

   Ans. b
   p. 112

23. Select the <u>correct</u> statement.
  a. Some drugs readily pass through the skin.
  b. Drugs given by inhalation are usually slower acting and less potent than drugs given orally.
  c. Usually more drug is needed when given I.V. than when administered orally.
  d. The more times a drug is injected in the same vein, the easier the injection becomes.

   Ans. a
   p. 112

24. Select the <u>incorrect</u> statement.
  a. Drugs that are water soluble are not likely to get into organs with a high fat content.
  b. Drugs that are not very fat soluble are most likely to affect the brain.
  c. The principal factor that determines if drugs will cross the placental barrier is molecular size.
  d. Most drugs are able to pass across the placental barrier.

   Ans. b
   p. 113

25. You have discovered that you require at least 300 mg of aspirin to get any pain relief. Which of the following terms best describes this phenomenon?
    a. the plateau dose
    b. the cumulative effect
    c. the biological half-life
    d. the threshold dose

    Ans. d
    p. 113

26. Selective filtering between the cerebral blood vessels and the brain is called a _____.
    a. cumulative effect
    b. blood-brain barrier
    c. biotransformation
    d. therapeutic index

    Ans. b
    p. 113

27. _____ is the maximum drug effect, regardless of dose.
    a. Threshold
    b. Cumulative effect
    c. Plateau effect
    d. Metabolism

    Ans. c
    p. 114

28. Chemical alteration of drugs by body processes is called _____.
    a. metabolism
    b. threshold
    c. teratogenic
    d. none of the above

    Ans. a
    p. 114

29. Select the incorrect statement concerning metabolism.
    a. It usually inactivates drugs.
    b. It usually helps to eliminate drugs more rapidly.
    c. Hormones and neurotransmitters are inactivated by metabolism.
    d. The rate of drug metabolism always stays constant.

    Ans. d
    p. 114-115

30. Select the correct statement.
    a. If metabolizing liver enzymes are induced, tolerance to the effects of drugs may occur.
    b. Most drugs and metabolites are excreted from the body in the feces.
    c. Water-soluble drugs are not excreted by the kidneys.
    d. Making the urine more acidic does not influence the excretion of drugs.

    Ans. a
    p. 115

31. Select the <u>incorrect</u> statement.   Ans. b
    a. Usually the elderly should be given   p. 115-
       lower doses than other adults to avoid   116
       adverse effects.
    b. There are usually substantial differences
       in the way males and females respond to
       drugs.
    c. The condition of pregnancy increases a
       women's vulnerability to drug side effects.
    d. A history of hepatitis often results in a
       longer duration of drug effects and increased
       likelihood of side effects.

32. _____ is a form of withdrawal or a paradoxical   Ans. b
    effect that occurs when a drug has been   p. 116
    eliminated from the body of a person who has
    become physically dependent.
    a. Physical dependence
    b. Rebound effect
    c. Reverse tolerance
    d. Mental set

33. _____ refers to changes causing decreased   Ans. d
    response to a set dose of a drug.   p. 117
    a. Potency
    b. Toxicity
    c. Metabolism
    d. Tolerance

34. Continual use of barbiturates stimulate   Ans. c
    formation of metabolizing enzymes that   p. 118
    inactivate these drugs. Thus, higher doses
    of barbiturates are required to get an effect.
    This phenomenon <u>is described</u> by which of the
    following terms?
    a. behavioral compensation
    b. pharmacodynamic tolerance
    c. disposition tolerance
    d. reverse tolerance

35. Select the <u>incorrect</u> statement.   Ans. d
    a. Physical dependence is a physiological   p. 119-
       state of adaptation to a drug.   120
    b. Physical dependence can result in withdrawal
       symptoms when the use of the drug is
       abruptly stopped.
    c. Rebound effects are a form of withdrawal.
    d. Rebound effects usually mimic the effects of
       the drugs themselves.

36. Select the correct statement. Ans. a
    a. Use of cocaine can cause sensitization. p. 119
    b. Physical dependence almost always occurs with chronic use of the hallucinogens.
    c. Intense use of CNS depressants does not cause physical dependence.
    d. Physical dependence caused by chronic narcotic use usually profoundly disrupts normal daily activities and personal interactions.

37. Select the incorrect statement concerning psychological dependence. Ans. b
    p. 120
    a. It often occurs from using drugs with abuse potential.
    b. It causes severe physical discomfort and rebound when drug use is stopped.
    c. It can cause intense craving and lure former drug abusers back to their drug habits.
    d. It can occur independent of physical dependence.

38. Select the incorrect statement concerning the "placebos". Ans. c
    p. 121
    a. They possess no pharmacological action.
    b. Placebos often work well in the treatment of pain.
    c. Placebos are usually more beneficial when patient-doctor relationships are cold and aloof.
    d. Placebos may work by activating endogenous systems such as the endorphin peptides.

39. Select the incorrect statement concerning factors that contribute to drug abuse patterns. Ans. a
    p. 122-124
    a. Peer pressure usually is not an important factor contributing to initial drug experimentation.
    b. Many persons who abuse drugs are attempting to self-medicate some form of mental disorder.
    c. Cocaine is frequently self-administered used to treat depression.
    d. A useful treatment for cocaine abuse is the antidepressant desipramine.

ESSAYS

1. What is the difference between potency and toxicity? Give examples of each.

2. List and briefly discuss the pharmacokinetic factors that can influence the effects caused by drugs.

3. Discuss how pharmacokinetic factors can cause people to have different responses to the same drug.

4. What does fat solubility have to do with drug use and abuse? How does this affect the actions of LSD?

5. Define and explain the following: tolerance, withdrawal, rebound, physical dependence, and psychological dependence.

6. Should placebos be given to patients who tend to over medicate themselves? Why or why not? Defend your answer from a medical as well as an ethical standpoint.

SUPPLEMENTARY MEDIA

ALMOST EVERYONE DOES. This fourteen minute 16mm film probes drug abuse from aspirin to heroin relating its cause to the desire to relieve unpleasant feelings. Observes the effect of advertising, smoking, and drinking on a young boy with normal emotional difficulties, and points out that drugs act as a temporary escape. Argues that personal involvement is needed to overcome depressive moods. Available from Indiana University Audio-visual Center, Bloomington, Indiana 47405-5901.

AMERICA: HOOKED ON DRUGS. This twenty-two minute video explores the prevalence of drug use in this country and the debilitating effects it has on the human brain, as well as the personal costs and the loss to business in declining productivity. Features interviews with three former drug addicts who speak candidly about their experiences. Emphasizes the need for anti-drug education beginning at the elementary school level and points out the governmental and community efforts to combat the epidemic. Available from the Indiana University Audio-visual Library, Bloomington, Indiana 47405-5901.

POWER OF ADDICTION (GX 1976-VHS). This nineteen minute video discusses both chemical and behavioral addiction and analyzes their causes. Available from "Films for the Humanities & Sciences," P.O.Box 2053, Princeton, NJ 08543, or call (800) 257-5126.

– Chapter 5 –
HOMEOSTATIC SYSTEMS AND DRUGS

This chapter discusses those systems that are responsible for maintaining the body's internal stability or homeostasis particularly as they are affected by substances of abuse. The continual physiological adjustments that are necessary to optimize body functions are controlled by the release of endogenous chemicals either from the nervous systems (neurotransmitters) or the endocrine system (hormones) that act at target sites on membranes called receptors. Because all drugs of abuse have profound effects on the body's homeostasis, the text introduces the students to basic elements of the nervous system, followed by an examination of its major divisions: the central, autonomic and peripheral neurons systems. Examples of how and why substances of abuse influence the activity of these nervous systems are presented. The components and operation of the endocrine system are also discussed in specific as they relate to drug effects. The use of anabolic steroids profoundly influences the endocrine systems and is dealt with at length.

TRUE OR FALSE

1. Some natural chemicals, produced by the body, have the same effect as narcotic drugs. — Ans. T p. 129

2. Drugs that affect the neurotransmitter dopamine usually alter both mental state and motor activity. — Ans. T p. 129

3. The anabolic steroids often abused by athletes are chemically related to female hormones. — Ans. F p. 129

4. Anabolic steroids are considered non-controlled drugs by the Drug Enforcement Administration (DEA). — Ans. F p. 129

5. The two principal systems that help human beings maintain homeostasis are the nervous system and the endocrine system. — Ans. T p. 130

6. Neurotransmitters are chemical messengers that are released from the terminals of neurons. — Ans. T p. 131

7. Neurons usually communicate with each other by releasing hormones. — Ans. F p. 131

8. Hormones tend to have a faster onset of action than do the neurotransmitters. — Ans. F p. 132

9. Hormones are regulatory chemicals released by endocrine systems. — Ans. T p. 132

10. The building block of the nervous system is the neuron.   Ans. T p. 133

11. Short branches of neurons that receive transmitter signals are called receptors.   Ans. F p. 134

12. An axon is an extension of the neuronal cell body along which electrochemical signals travel.   Ans. T p.135

13. An inhibitory synapse usually increases the likelihood of neurotransmitters being released.   Ans. F p. 136

14. The substance L-DOPA is converted into serotonin in the brain.   Ans. F p. 138

15. The acetylcholine sites have been divided into two main subtypes based on the response to two drugs.   Ans. T p. 138

16. Norepinephrine and epinephrine receptors are separated into the categories of alpha and beta.   Ans. T p. 139

17. Dopamine is an important transmitter in controlling movement and fine-muscle activity.   Ans. T p. 140

18. The central nervous system consists of the brain and the spinal cord.   Ans. T p. 140

19. LSD likely exerts its hallucinogenic effects by nicotinic receptors.   Ans. F p. 140

20. Drugs that increase dopamine activity usually affect both motor activity and mental states.   Ans. T p. 140

21. The limbic system in the brain helps to regulate emotional activities and memory.   Ans. T p. 142

22. Laboratory animals find the administration of stimulants of abuse into limbic structures of the brain unpleasant.   Ans. F p. 143

23. The sympathetic and parasympathetic nervous systems usually have the same effect on organs.   Ans. F p. 144

24. The sympathetic nervous system uses dopamine as a transmitter.   Ans. F p. 144

25. The parasympathetic uses acetylcholine as a transmitter.   Ans. T p. 144

26. The Autonomic Nervous System is divided into two functional components, the sympathetic and the parasympathetic nervous systems.   Ans. T p. 144

27. The pituitary gland is often referred to as the master gland.
Ans. T
p. 146

28. The long term regular use of anabolic steroids has not been thoroughly studied.
Ans. T
p. 147

29. Anabolic steroids are compounds chemically like the steroids that stimulate production of tissue mass.
Ans. T
p. 147

30. Anabolic steroids have been proven to dramatically improve athletic performance in adult men.
Ans. F
p. 147

31. Anabolic steroid abuse is only a significant problem with athletes participating in organized or professional sports.
Ans. F
p. 147

32. Steroids can cause aggressive behavior and excitation.
Ans. T
p. 148

MULTIPLE CHOICE

1. _____ is the maintenance of internal stability.
   a. Dependence
   b. Withdrawal
   c. Homeostasis
   d. Rebound

   Ans. c
   p. 130

2. _____ are the principal cells in nervous systems.
   a. neurons
   b. hormones
   c. endorphins
   d. blood

   Ans. a
   p. 130

3. Select the <u>incorrect</u> statement.
   a. The endocrine system helps to maintain the homeostasis of the body by releasing neurotransmitters.
   b. Hormones are usually released into the bloodstream and carried by the blood to all the organs and tissues of the body.
   c. c.AMP is a secondary messenger chemical in cells that is often affected by hormones.
   d. Neurotransmitters cause either increased or decreased activity of organs.

   Ans. a
   p. 130-133

4. _____ are chemical messengers released by neurons.
   a. Dendrites
   b. Receptors
   c. Neurotransmitters
   d. Hormones

   Ans. c
   p. 131

5. Neurotransmitters that have narcotic-like effects are called _____.
    a. endorphins
    b. dopamine
    c. GABA
    d. nonadrenaline

    Ans. a
    p. 132

6. What system sets the limits for proper functioning of the nervous system?
    a. vascular
    b. endocrine
    c. immune
    d. postsynaptic

    Ans. b
    p. 132

7. The nervous system is composed of all but which of the following?
    a. the brain
    b. the spinal cord
    c. the neurons
    d. the blood vessels

    Ans. d
    p. 133

8. How many neurons are in the brain?
    a. under 1 billion
    b. 1-3 billion
    c. 5 billion
    d. over 10 billion

    Ans. d
    p. 134

9. A minute gap between the neuron and target cell across which neurotransmitters travel is called _____.
    a. dendrite
    b. synapse
    c. receptor
    d. neuron

    Ans. b
    p. 134

10. Which of the following list describes a neuronal axon?
    a. they are protein sites on cells that react to endogenous chemical messengers such as neurotransmitter
    b. a small fiber that conducts electric-like impulses from dendrites to terminal
    c. part of the neuron's cell body
    d. short tree-like branches that pick up information from surrounding neurons.

    Ans. b
    p. 135

11. Select the <u>incorrect</u> statement concerning receptors in the nervous system.
    a. They are protein sites on cells that respond to endogenous chemical messengers, such as transmitters.
    b. Activation of most receptors alters the functioning of the cell.
    c. Drugs that are agonists keep the receptor from being activated.
    d. Many psychoactive drugs exert their pharmacological effects by either activating or inactivating receptors on neurons.

    Ans. c
    p. 137

12. A class of biochemical compounds including no-repinephrine and dopamine is called _____.
    a. sympathomimetic
    b. catecholamines
    c. synapse
    d. muscarinic

    Ans. b
    p. 138

13. Muscarinic receptors are activated by which of the following transmitters?
    a. acetylcholine
    b. GABA
    c. substance P
    d. dopamine

    Ans. a
    p. 138

14. Which of the following neurotransmitters is <u>not</u> classified as a catecholamine?
    a. acetylcholine
    b. dopamine
    c. norepinephrine
    d. epinephrine

    Ans. a
    p. 138-139

15. Which of the following neurotransmitters is metabolized by the enzyme, monoamine oxidase (MAD)?
    a. acetylcholine
    b. GABA
    c. substance P
    d. dopamine

    Ans. d
    p. 138-139

16. Which of the following is <u>not</u> a neurotransmitter?
    a. corticosteroids
    b. acetylcholine
    c. dopamine
    d. serotonin

    Ans. a
    p. 138-140

17. Select the <u>incorrect</u> statement. Ans. d
    a. The Reticular Activating System helps to p. 139
       regulate the brain's state of arousal. 140
    b. Parkinson's disease is due to damage of
       dopamine neurons in the basal ganglia.
    c. Drugs that alter the activity of dopamine-
       releasing neurons often affect motor activity.
    d. For the most part, drugs of abuse decrease the
       activity of dopamine in the limbic system.

18. How much does the average brain weigh? Ans. c
    a. 1/2 pound p. 140
    b. 1 pound
    c. 3 pounds
    d. 4 1/2 pounds

19. LSD is thought to do which of the following? Ans. a
    a. have serotonin-like chemical effects p. 140
    b. cause a deficiency of dopamine in the basal
       ganglia of the brain
    c. cause a deficiency of norepinephrine in the
       cortex
    d. cause a deficiency of acetylcholine in the
       basal ganglia of the brain

20. Agents that antagonize the effects of acetyl- Ans. a
    choline are called _____. p. 142
    a. anticholinergic
    b. anabolic
    c. testosterones
    d. narcotics

21. Which of the following is <u>not</u> part of the nervous Ans. d
    system? p. 142-
    a. hypothalamus 143
    b. limbic system
    c. basal ganglia
    d. pancreas

22. What part of the brain has developed most in Ans. d
    the evolutionary process? p. 143
    a. basal ganglia
    b. limbic
    c. hypothalamus
    d. cortex

23. Which of the following brain regions is <u>most</u> Ans. a
    <u>likely</u> to directly regulate the endocrine systems? p. 143
    a. hypothalamus
    b. basal ganglia
    c. cortex
    d. cerebellum

24. Which area of the cortex is the most developed in man compared to lower animals?
    a. sensory areas
    b. association areas
    c. receiving areas
    d. output areas

    Ans. b
    p. 143

25. Select the incorrect statement concerning the hypothalamus.
    a. It serves as the CNS control center for the autonomic nervous system.
    b. It is important in the CNS regulation of endocrine systems.
    c. Because it is isolated, drugs are less likely to affect the hypothalamus than other brain areas.
    d. It is important in maintaining the homeostasis of the body.

    Ans. c
    p. 143

26. Which of the following effects is least likely to be caused by a sympathomimetic drug?
    a. raises blood pressure
    b. slows heart rate
    c. reduces muscle contraction in the stomach
    d. enlarges the pupils of the eye

    Ans. b
    p. 144

27. Which of the following is not a major nervous system?
    a. sympathetic
    b. autonomic
    c. parasymphathetic
    d. adrenal

    Ans. d
    p. 144-146

28. Which is often referred to as the master gland?
    a. tear gland
    b. brain
    c. pituitary gland
    d. none of the above

    Ans. c
    p. 146

29. Which of the following is most similar to the anabolic steroids often abused by athletes?
    a. estrogens
    b. prostoglandins
    c. adrenalin
    d. testosterone

    Ans. d
    p.146

30. Select the <u>incorrect</u> statement concerning the adrenal glands.
    a. The corticosteroids are released from the adrenal cortex.
    b. Adrenaline is released from the adrenal medulla.
    c. Small amounts of androgens are released by the adrenal cortex.
    d. Adrenaline is classified as an anabolic steroid.

    Ans. d
    p. 146

31. Select the <u>correct</u> statement.
    a. Testosterone is released from the ovaries and causes female sex changes.
    b. Sex hormones can affect emotional states.
    c. Most drugs used to treat endocrine problems are given to replaced efficient hormones.
    d. Diabetes is usually due to a lack of insulin secreted by the pancreas.

    Ans. b
    p. 146

32. Chemicals that are able to convert nutrients into tissue mass are called _____.
    a. anabolic chemicals
    b. anticholinergic chemicals
    c. vasculature chemicals
    d. neuronal chemicals

    Ans. a
    p. 147

33. Which of the following is <u>least likely</u> to be caused by abusing anabolic steroids?
    a. stimulate the growth of body hair in females
    b. cause sedation and a mellowing of temper
    c. deepening of the voice in women
    d. increased acne

    Ans. b
    p. 147-148

34. Anabolic steroids have been classified in which of the following Schedules of the Controlled Substance Act?
    a. I
    b. II
    c. III
    d. IV

    Ans. c
    p. 148

35. Anyone who distributes anabolic steroids are required by law to be registered with which of the following?
    a. FDA
    b. FBI
    c. DEA
    d. CIA

    Ans. c
    p. 148

ESSAYS

1. Explain the similarities and differences between the nervous

system and the endocrine system.

2. Describe the functions of a neuron and how drugs of abuse can influence a neuron's activity.

3. Identify which brain areas are most likely to be affected by drugs of abuse.

4. Explain how and why anabolic steroids are abused and the social impact of that abuse.

SUPPLEMENTARY MEDIA

DRUG PROFILES: THE PHYSICAL AND MENTAL ASPECTS. This twenty-eight minute video provides detailed information about ten drugs of abuse: cocaine, heroin, methaqualone, alcohol, marijuana, barbiturates, amphetamines, tranquilizers, PCP, and LSD. Suggests that because substance abuse is such a large problem, employers, supervisors, educators, and others in leadership roles must be able to recognize the signs and symptoms of abuse. Gives special attention to drugs and pregnancy and drugs and driving. Available from Indiana University Audio-visual Library, Bloomington, Indiana 47405-5901.

DRUGS AND THE NERVOUS SYSTEM (SECOND REVISION). This eighteen minute 16mm film explains, through animation, how drugs affect the nervous system and the resulting changes in bodily and reflex behaviors and perceptions. Follows the effect of aspirin to reduce pain and fever. Discusses fumes from glue, depressants, alcohol, heroin, stimulants, marijuana, LSD, and PCP. Available from Indiana University Audio-visual Library, Bloomington, Indiana 47505-5901.

THE ADDICTED BRAIN(GX 1363-VHS). This twenty-six minute video explains the brain chemistry that is the basis for addiction and addictive behavior. Available from "Films for the Humanities & Sciences," P.O.Box 2053, Princeton, NJ 08543 or call (800) 257-5126.

DRUG ABUSE AND THE BRAIN(1993). This twenty minute video provides a detailed look at the biological basis of drug addiction. Available from the National Clearinghouse for Alcohol and Drug Information (NCADI), P.O.Box 2345, Rockville, MD 20847-2345 or call (800)729-6686.

– Chapter 6 –
CNS Depressants
Sedative-Hypnotics

Central nervous system (CNS) depressants are some of the most widely used and abused drugs in this country. Chapter 6 explains the history of these drugs and why they are so appealing and dangerous. The unique clinical and abuse features of these agents are also discussed. The three main groups of CNS depressants dealt with in this chapter are (1) the very popular benzodiazepines (e.g., Valium and Halcion), (2) the waning barbiturates (e.g. seconal) and (3) the miscellaneous depressants, (e.g., chloral hydrate and methaqualone) most of which have barbiturate-like properties. The chapter compares the abuse potential, side effects and clinical usefulness of the different groups of depressants. Finally, chapter six discusses the abuse patterns of the depressants and the unique problems of treating dependence on, and withdrawal from, these drugs.

TRUE OR FALSE

1. Low doses of alcohol can cause sedative effects and relaxation.
   Ans. T
   p. 154

2. Many individuals who are dependent on depressants acquire them byy prescription.
   Ans. T
   p. 154

3. During the 1970s and 1980s, approximately twice as many women took depressants as men.
   Ans. T
   p. 156

4. The effects of the CNS depressants tend to be dose dependent; therefore, a high dose of sedative has a hypnotic effect.
   Ans. T
   p. 156

5. Amnesia is a state of deep depression of the CNS, and is characterized by controlled unconsciousness.
   Ans. F
   p. 156

6. CNS depressants used as hypnotics are intended to help induce sleep.
   Ans. T
   p. 156

7. In high doses, some CNS depressants can cause anesthesia.
   Ans. T
   p. 156

8. Benzodiazepines and barbiturates are sometimes used to treat seizures.
   Ans. T
   p. 157

9. A person who is physically dependent on barbiturates can experience seizures if drug use is stopped abruptly.
   Ans. T
   p. 157

10. Barbiturates are safer drugs than the benzodiazepines.
Ans. F
p. 157

11. Short-acting depressants are better suited to treating persistent problems such as anxiety and stress.
Ans. F
p. 158

12. The presence of benzodiazepines increases the inhibitory effects of GABA.
Ans. T
p. 158

13. Benzodiazepines do not significantly interact with other CNS depressants.
Ans. F
p. 159

14. Some side effects of benzodiazepine-type drugs are drowsiness, nausea, and diminished libido.
Ans. T
p. 159

15. On rare occasions, benzodiazepines can produce paradoxical effects such as anxiety, or bizarre behavior.
Ans. T
p. 159

16. Prolonged use of hypnotic doses of benzodiazepines can cause insomnia when the drug is stopped.
Ans. T
p. 160

17. Most people that become dependent on the benzodiazepines get them illegally.
Ans. F
p. 161

18. It usually is better to rapidly withdraw benzodiazepines from persons who have become physically dependent on these drugs.
Ans. F
p. 161

19. Patients using barbiturates to produce sleep usually wake up feeling rested and relaxed.
Ans. F
p. 162

20. After regular use, stopping the use of barbiturates abruptly can cause life-threatening seizures.
Ans. T
p. 163

21. The CNS depressants are classified as Schedule I, Schedule II, Schedule III, and Schedule IV according to their relative potential for physical and psychological dependence and their perceived clinical usefulness.
Ans. T
p. 163

22. The clinical use of barbiturates has increased dramatically in recent years.
Ans. F
p. 164

23. Barbiturates can distort mood and impair judgement and motor skills even the day after use.
Ans. T
p. 164

24. Pharmacodynamic tolerance develops over weeks or months, where as drug-disposition tolerance peaks in a few days.
Ans. T
p. 164

25. Liver damage increases the likelihood of experiencing dangerous side effects to the barbiturates.
Ans. T
p. 164

26. Barbiturates appear to act on specific receptor sites, but benzodiazepines do not.
Ans. F
p. 164

27. The fat-solubility of the barbiturates is important in determining their duration of action.
Ans. T
p. 164

28. Antihistamines are considered relatively safe drugs.
Ans. T
p. 167

29. Dependence or abuse rarely develop when using antihistamines.
Ans. T
p. 167

30. Sympathomimetics are usually the active ingredient included in OTC sleep aids.
Ans. F
p. 167

31. Antihistamines are sometimes used as sedatives.
Ans. T
p. 167

32. Methaqualone (Quaalude) is classified as a Schedule IV Controlled Substance.
Ans. F
p. 167

33. The antihistamine, diphenhydramine, is commonly used as an OTC sleep aid.
Ans. T
p. 167

34. All sedative-hypnotics can produce severe physical dependence.
Ans. T
p. 168

35. The longer-acting CNS depressants are the most likely to be abused.
Ans. F
p. 169

36. Pentobarbital is often used in the treatment of severe dependence on sedative-hypnotics.
Ans. T
p. 170

37. Withdrawal from long-acting CNS depressants is more intense than withdrawal from short-acting depressants.
Ans. F
p. 170

MULTIPLE CHOICE

1. Barbiturates are _____ valium-like benzodiazepines.
   a. safer than
   b. as safe as
   c. more dangerous than
   d. none of the above

Ans. c
p. 153

2. Select the <u>incorrect</u> statement. Ans. c
   a. Doses of barbiturates that are required p. 155
      to relieve anxiety can also suppress
      respiration.
   b. Benzodiazepines are relatively safe when used
      for short periods of time.
   c. Barbiturates are prescribed for the relief of
      anxiety more than the benzodiazepines.
   d. The prescribing of depressants has declined
      over the last 10 years.

3. A CNS depressant can _____. Ans. d
   a. relieve anxiety p. 156
   b. be used as a sleep aid
   c. cause severe physical dependence
   d. all of the above

4. Depressants are often classified according to Ans. c
   _____. p. 156
   a. their MSA rating
   b. the degree to which they react with alcohol
   c. the degree they decrease CNS activity
   d. the first letter in their names

5. A CNS depressant often used to relieve anxiety Ans. b
   is a (an) _____. p. 156
   a. hypnotic
   b. sedative
   c. valium
   d. hypnotic

6. Which of the following depressants is named Ans. d
   after a Greek god? p. 156
   a. amnesiacs
   b. sedatives
   c. valium
   d. hypnotics

7. Which of the following terms refers to drugs Ans. b
   used to relieve anxiety? p. 156
   a. hypnotic
   b. anxiolytic
   c. analgesic
   d. antipsychotic

8. Which of the following statements concerning Ans. b
   benzodiazepines is <u>incorrect</u>? p. 157-
   a. There are currently more than 10 different 162
      benzodiazepines marketed in this country.
   b. Therapeutic doses of the benzodiazepines
      depress respiration.
   c. Benzodiazepines have less effect on REM
      sleep than barbiturates.
   d. Frequent use of the benzodiazepines can
      cause tolerance and withdrawal.

9. Select the correct statement.   Ans. a
   a. Xanax and Halcion are popular benzodiaz-   p. 157-
      epines.   158
   b. Benzodiazepines are very similar to the
      tranquilizers.
   c. Librium and Valium are barbiturates.
   d. Because of its problems, Valium is rarely
      prescribed today.

10. Benzodiazepines are classified in which of the   Ans. d
    following Schedules?   p. 158
    a. I
    b. II
    c. III
    d. IV

11. Benzodiazepines are frequently used clinically   Ans. d
    to _____.   p. 158
    a. aid in withdrawal from alcohol
    b. treat seizures
    c. preanesthetize a patient
    d. all of the above

12. For most of the 1970s, _____ was the top-selling   Ans. c
    prescription drug.   p. 158
    a. Clonazepam
    b. Librium
    c. Valium
    d. Xanax

13. Benzodiazepines are usually classified as _____   Ans. c
    a. Schedule I   p. 158
    b. Schedule II
    c. Schedule IV
    d. Schedule V

14. Approximately _____ benzodiazepine compounds are   Ans. b
    available in the United States currently.   p. 158
    a. 5
    b. 12
    c. 25
    d. 49

15. Select the incorrect statement.   Ans. a
    a. Narcotic addicts rarely combine benzodi-   p. 159-
       azepines with heroin.   161
    b. Sometimes benzodiazepines are combined
       with cocaine to reduce the "crash" after
       stimulant use.
    c. Some benzodiazepines can cause paradoxical
       behavior such as nightmares, instability
       and restlessness.
    d. Chronic use of benzodiazepines does not
       appear to cause permanent neurological
       damage.

16. Frequent use of benzodiazepines can lead to _____.
    a. withdrawal
    b. dependence
    c. tolerance
    d. all of the above

    Ans. d
    p. 160

17. Benzodiazepines are relatively safe when used for _____.
    a. short periods of time
    b. long periods of time
    c. any duration of time
    d. none of the above

    Ans. a
    p. 161

18. Barbituric acid was synthesized in 1864 by _____.
    a. A. Bayer
    b. W. Excedrin
    c. J. Salk
    d. I. Westermann

    Ans. a
    p. 162

19. Which of the following usually causes death resulting from a barbiturate overdose.
    a. respiratory or cardiovascular depression
    b. seizures
    c. hemorrhaging
    d. heart attack or stroke

    Ans. a
    p. 162

20. Short-acting barbiturates, such as secobarbital are usually classified in which of the following Schedules?
    a. I
    b. II
    c. III
    d. IV

    Ans. b
    p. 163

21. Short-acting barbiturates (like secobarbital) are classified as _____.
    a. Schedule I
    b. Schedule II
    c. Schedule III
    d. Schedule IV

    Ans. b
    p. 163

22. _____, an ultrashort acting barbiturate, is often used as anesthesia for minor surgery.
    a. Amytal
    b. Luminal
    c. Pentothal
    d. Butisol

    Ans. c
    p. 164

23. Barbiturates are generally classified in terms of _____.
    a. their duration of action
    b. the first letter in their name
    c. the degree to which they react to alcohol
    d. their ability to cause excitation

    Ans. a
    p. 164

24. Women's reaction to barbiturates is most likely to be different than men's, due to differences in _____.
    a. hormones
    b. pH balance
    c. height
    d. body-fat ratio

    Ans. d
    p. 164

25. Which would result from enzyme induction in the liver?
    a. sensitization
    b. tolerance
    c. withdrawal
    d. allergies

    Ans. b
    p. 164

26. Select the incorrect statement concerning barbiturates.
    a. Physical dependence usually occurs within one week of using the barbiturates.
    b. Women and men may respond differently to the barbiturates because of the difference in the fat content of their bodies.
    c. Repeated use of the barbiturates at short intervals can cause induction of metabolizing enzymes in the liver.
    d. Barbiturates appear to increase the activity of the neurotransmitter, GABA.

    Ans. a
    p. 164-165

27. Phenobarbital is often used to treat which of the following conditions?
    a. migraine headaches
    b. ulcers
    c. pain
    d. seizures

    Ans. d
    p. 165

28. The nickname "_____" refers to the drug pentobarbital.
    a. blue heaven
    b. double-trouble
    c. purple heart
    d. yellow jacket

    Ans. d
    p. 165

29. A "Mickey Finn" was used to make a person temporarily unconscious (knock-out) and contained _____.
    a. methyprylon
    b. glutethimide
    c. chloral hydrate
    d. ethchlorvynol

    Ans. c
    p. 166

30. Drugs often used to relieve symptoms of allergies and motion sickness are called _____.
    a. antihistamines
    b. anticholinergics
    c. methyprylones
    d. methaqualones

    Ans. a
    p. 167

31. Select the <u>incorrect</u> statement about methaqualone.
    a. Methaqualone was used as "knock-out drops" on Barbary Coast of San Francisco and when mixed with alcohol was called a "Mickey Finn".
    b. Methaqualone has been used in the treatment of malaria.
    c. Methaqualone is no longer legally available in the United States.
    d. High doses can cause psychological and physical dependence like the barbiturates.

    Ans. a
    p. 167

32. Select the <u>correct</u> statement concerning antihistamines.
    a. Tolerance to antihistamine-induced sedation develops slowly.
    b. OTC antihistamines are frequently abused.
    c. Many antihistamines can cause annoying anticholinergic side effects.
    d. They do not significantly interact with other CNS depressants.

    Ans. c
    p. 167-8

33. The abuse of CNS depressants from 1991 to 1992 _____.
    a. increased
    b. declined dramatically
    c. remained the same
    d. unknown

    Ans. c
    p. 170

34. Withdrawal symptoms from _____ depressants are the most severe.
    a. legal
    b. illicit
    c. long-acting
    d. short-acting

    Ans. d
    p. 170

35. The approach to detoxifying a person dependent on drugs depends on _____.
    a. the nature of the drug
    b. the severity of dependence
    c. the duration of action of the drug
    d. all of the above

    Ans. d
    p. 170

ESSAYS

1. What are the principal dose dependent effects of CNS depressants in Figure 6-1, p. 156.

2. Why are CNS depressant drugs commonly abused?

3. Describe the differences in effects between short- and long-acting CNS depressants?

4. List and discuss the four different types of persons who abuse CNS depressants?

5. What are the principal ways that CNS depressants interact with other drugs?

6. Explain how prescribed CNS depressants typically become abused substances.

SUPPLEMENTARY MEDIA

FOCUS ON DOWNERS. This fourteen minute 16mm film reviews the characteristics of barbiturates or "downers" and their proper use pointing out that addiction takes only a few weeks and that withdrawal symptoms are severe. Presents vignettes which portray death due to an overdose, combining downers and alcohol, and driving while under the influence of drugs. Emphasizes that while many users are trying to overcome depression caused by school, home, or environmental issues, these problems cannot be solved by using drugs. Available from Indiana University Audio-visual library, Bloomington, Indiana 47405-5901.

SEDATIVES. The potential physiological and psychological dangers of this class of drugs is discussed as well as the medical usefulness. Available from the University of Maine Instructional Systems Center, 12 Shibles Hall, Orono, Maine 04469.

— Chapter 7 —
Alcohol: Pharmacological Effects

The pervasive effects of alcohol use are covered in this chapter and the next. This chapter addresses a largely pharmacological explanation of alcohol as a drug. Alcohol (more precisely termed ethanol) is briefly discussed as a drug and as a social drug. Following this, the chapter delves into the nature and history of alcohol, the properties and types of alcohol, such s methyl alcohol (wood alcohol) ethylene glycol (used as an antifreeze), propyl alcohol (commonly known as rubbing alcohol) and ethanol (alcohol used for drinking purposes). (Note that the first three are poisonous.) Next, the physical effects of alcohol; alcohol and tolerance; experiencing the effects of and metabolizing alcohol; polydrug use; short- and long-term effects of alcohol use are presented.

The last section elaborates on the effects of alcohol on the organ systems and bodily functions of: the brain and nervous system; liver; digestive system; blood, cardiovascular system; sexual organs; endocrine's; kidneys; mental disorder and change to the brain; fetus; and a short discussion on fetal alcohol syndrome (FAS).

TRUE OR FALSE

1. The liver metabolizes alcohol at a rate proportional to the amount ingested. — Ans. F p. 175

2. Alcohol is a psychoactive substance because it affects both brain and body functions. — Ans. T p. 176

3. Decreased life expectancy, increased homicides and suicides, and increased child abuse are some of the costs of problem drinking and alcoholism in the United States. — Ans. T p. 177

4. At sometime in their lives, almost 80 percent of all Americans will be involved in an alcohol-related traffic accident. — Ans. F p. 177

5. Alcohol is the leading cause of premature death in America. — Ans. F p. 177

6. The process of making alcohol, called fermentation is a synthetic one. — Ans. F p. 177

7. Five percent is the natural limit of alcohol found in fermented wines or beers. — Ans. F p. 177

8. The invention of the still has greatly increased the problem of alcohol abuse. — Ans. T p. 177

9. Approximately half of the deaths caused by traffic accidents in the United States are alcohol related. Ans. T p. 177

10. Alcohol is a chemical structure that has hydroxylgroup attached to a carbon atom. Ans. T p. 178

11. Ninety proof whiskey has a 45 percent alcohol content. Ans. T p. 179

12. Congeners are the nonalcoholic constituents of alcoholic beverages. Ans. T p. 179

13. High concentrations of congeners speed up alcohol absorption. Ans. F p. 179

14. An anesthetic is a drug that produces sensitivity to pain. Ans. F p. 179

15. The effects of alcohol on the human body depends on the amount of alcohol in the brain or BAC. Ans. F p. 179

16. Generally the more alcohol in the stomach the greater the absorption rate. Ans. T p. 180

17. The tolerance developed to alcohol is comparable to that of barbiturates. Ans. T p. 181

18. Behavioral Tolerance is due to changes in the matter in which alcohol effects the body. Ans. F p. 181

19. For the effects of drugs to be experienced they must reach the brain. Ans. T p. 181

20. Alcohol is actually an effective bactericide, especially when used on open wounds. Ans. F p. 182

21. Using other types of drugs with alcohol is called Polydrug use. Ans. T p. 182

22. Coffee is not effective in "sobering up" the individual with a hangover. Ans. T p. 183

23. The symptoms of a hangover are usually most severe an hour after drinking, and diminish as the alcohol leaves the body. Ans. F p. 183

24. Although an alcoholic may metabolize alcohol more rapidly than a non-alcoholic, the alcohol toxicity level of alcohol stays the same. Ans. T p. 183

25. Every part of the brain and nervous system is affected and can be damaged by alcohol. Ans. T p. 185

26. Among alcoholics, intestinal disorders are the most common causes of death.
    Ans. F
    p. 185

27. The third stage of alcohol-induced liver disease is not reversible.
    Ans. T
    p. 186

28. Alcohol diminishes the effective functioning of the hematopoietic system.
    Ans. T
    p. 187

29. There is no safe amount of alcohol a pregnant woman can consume.
    Ans. T
    p. 189

MULTIPLE CHOICE

1. The consumable type of alcohol that is the psychoactive ingredient in alcoholic beverages is _____.
   a. Isopropanol
   b. Ethanol
   c. Ethylene glycol
   d. Methanol

   Ans. b
   p. 176

2. _____ is the process of making alcohol.
   a. Fermentation
   b. Liquification
   c. Pasteurization
   d. Distillation

   Ans. a
   p. 177

3. _____ is the process by which alcohol concentrations are increased.
   a. Liquification
   b. Fermentation
   c. Distillation
   d. Pasteurization

   Ans. c
   p. 177

4. The distillation devise or "still" was developed by the _____.
   a. Greeks
   b. Arabs
   c. English
   d. Chinese

   Ans. b
   p. 177

5. _____ is the process by which alcohol concentrations are increased.
   a. Liquification
   b. Fermentation
   c. Distillation
   d. Pasteurization

   Ans. c
   p. 177

6. _____ alcohol is used as an antifreeze, where as _____ alcohol is used as an antiseptic.
    a. Methyl, isopropyl
    b. Isopropyl, methyl
    c. Ethylene, isopropyl
    d. Isopropyl, ethylene

    Ans. c
    p. 178

7. Which of the following types of alcohol is <u>not</u> poisonous to humans?
    a. isopropyl alcohol
    b. ethyl alcohol
    c. methyl alcohol
    d. ethylene alcohol

    Ans. b
    p. 178

8. One hundred ten proof alcohol has a _____ alcohol content.
    a. 55 percent
    b. 50 percent
    c. 40 percent
    d. 45 percent

    Ans. a
    p. 179

9. The concentration of alcohol in table wine is usually about
    a. 1-5 percent
    b. 10-12 percent
    c. 15-25 percent
    d. 28-55 percent

    Ans. b
    p. 179

10. Drinks stronger than _____ proof may actually inhibit alcohol absorption.
    a. 70
    b. 80
    c. 90
    d. 100

    Ans. d
    p. 180

11. Which of the following people would have the highest blood alcohol level after consuming two glasses of wine?
    a. 100 lb. male
    b. 100 lb. female
    c. 150 lb. male
    d. 200 lb. male or female

    Ans. b
    p. 181

12. Alcoholic beverages contain large amounts of
    a. fats
    b. carbohydrates
    c. proteins
    d. minerals

    Ans. b
    p. 181

13. A glass of champagne has a faster onset of action than a glass of wine which contains the same amount of alcohol because
    a. champagne does not have to be digested because it is absorbed more readily, whereas wine has to be digested.
    b. the carbon dioxide (carbonation) in champagne moves it more rapidly from the stomach to the small intestine where absorption occurs more rapidly.
    c. liver enzymes do not metabolize champagne as rapidly as wine.
    d. wine, but not champagne, retains congeners which slow down absorption.

    Ans. b
    p. 181

14. As a rule it takes _____ the number of hours as the number of drinks consumed to sober up completely.
    a. equal
    b. half
    c. twice
    d. four times

    Ans. a
    p. 183

15. Alcohol withdrawal symptoms can include all except _____.
    a. insomnia
    b. anxiety
    c. spontaneous erections
    d. delirium

    Ans. c
    p. 183

16. The lethal blood alcohol level is _____.
    a. between 0.06% and 0.1%
    b. between 0.2% and 0.3%
    c. between 0.4% and 0.6%
    d. varies from state to state

    Ans. c
    p. 183

17. Which is the manner in which the body loses fluids through alcohol's diveretic action?
    a. perspiration
    b. urine
    c. saliva
    d. tears

    Ans. b
    p. 184

18. The "hair of the dog" method for treating a hangover consists of
    a. taking a drink of an alcoholic beverage.
    b. kissing the family dog.
    c. taking a drink of the same beverage that caused the hangover.
    d. taking an aspirin-caffeine combination to counteract the side effects.

    Ans. c
    p. 184

19. Prolonged heavy drinking can lead to <u>all</u> of the following except.
    a. kidney and liver damage
    b. scarring of the mucus lining
    c. lowered resistance to pneumonia and other infectious diseases
    d. damage to brain and peripheral nervous system

    Ans. b
    p. 185

20. About _____ of those individuals who chronically consume large amounts of alcohol suffer from cirrhosis.
    a. 5%
    b. 15%
    c. 30%
    d. 50%

    Ans. b
    p. 186

21. Cirrhosis of the liver _____.
    a. usually develops after about three years of heavy drinking
    b. is unique only to alcohol abusers
    c. is irreversible
    d. is the fourth leading cause of death in men aged 25 to 65 living in rural area

    Ans. c
    p. 186

22. Which of the following is not one of the three stages of alcohol-induced liver disease?
    a. cirrhosis of the liver
    b. hepatotoxic effect
    c. glycogen inflammation
    d. alcoholic hepatitis

    Ans. c
    p. 186

23. Regarding sexual performance and prolonged drug use, the findings suggest that alcohol
    a. has very little effect on sexual performance.
    b. can cause inflammation of the prostate gland.
    c. increases the sex drive.
    d. has less effort on women than men.

    Ans. b
    p. 188

24. Fetal alcohol syndrome is characterized by all of the following except:
    a. growth retardation before and/or after birth
    b. excessive weight gain
    c. a flattened bridge and shortened length of the nose
    d. abnormal features of the face and head

    Ans. b
    p. 189

25. Alcohol consumed during pregnancy can lead to what type of damage to newborn babies:
    a. spinal meningitis
    b. bacchanalia
    c. neuritis
    d. mental retardation

    Ans. d
    p. 189

ESSAYS

1. Describe the factors that affect the concentration of alcohol in the blood.

2. Explain how prolonged consumption of alcohol affects the cardiovascular system, liver and kidneys.

3. Define and explain congeners.

4. Identify the three types of poisonous alcohol and name the type that is used in alcoholic beverages.

5. What were the findings regarding alcohol use and (a)social class (b)religious (c)gender and (d)age differences?

SUPPLEMENTARY MEDIA

CODEPENDENTS. This thirty minute video presents dramatic vignettes played out from the point of view of Nancy Blaine, a codependent and wife of an alcoholic and trying to hold a family together in the face of this debilitating disease. Features two health professionals from Harvard Medical School who analyze Nancy's behavior in each dramatization, showing why alcoholism is a family disease. Available from Indiana University Audio-visual Library, Bloomington, Indiana 47405-5901.

CRUEL SPIRITS, ALCOHOL AND VIOLENCE. This thirty-two minute video documents the relationship between alcohol abuse and violence. Shows footage of community violence, emergency rooms, prisons, treatment centers, and police reports to confirm the connection between alcohol and violent incidents. Explores recent statistics and long term studies, incorporating interviews with medical personnel, researchers, police officials, prisoners, and treatment center residents. Examines alcohol as a factor for both victims and perpetrators of violence. Available from Indiana University Audio-visual Library, Bloomington, Indiana 47405-5901.

FETAL ALCOHOL SYNDROME. This thirteen minute 16mm film reveals the serious mental and physical defects often found in the children of mothers who drank heavily during pregnancy. Shows Kenneth, a 21-month-old boy, and his two school-age sisters who exhibit the defects, such as shortened eye slits, flat facial profile, small brain size, and learning impairments. Presents Dr. David Smith, professor of pediatrics at the University of Seattle, who has characterized the syndrome and has found that the effects caused during prenatal development are irreversible. Discusses rat research which supports Dr. Smith's theory and concludes with an interview with a mother whose drinking has seriously affected her child. Available from Indiana University Audio-visual Library, Bloomington, Indian 47405-5901.

ALCOHOL'S EFFECT ON THE BODY. This program illustrates the wide-ranging negative effects of alcohol on the human body and on society. Dr Ruth Liver delivers the facts on alcohol in the body; drugs and alcohol strain relationships on our soap parody. "The Young and the Breathless": and a fitness instructor describes how alcohol can ruin your appearance. (15 minutes, color) Films for the Humanities and Sciences Princeton, NJ.

ALCOHOL AND THE FAMILY: BREAKING THE CHAIN. This program analyzes the signs of alcoholism and shows how a family member, coworker, or friend can help break the chain; discusses the impact of alcoholism on the children of alcoholics; and evaluates the options and prognosis for alcoholism treatment. (25 minutes, color). Films for the Humanities and Sciences, Princeton, NJ.

CHILDREN OF ALCOHOLICS. Alcohol abuse means physical or psychological abuse of the family as well, which geometrically increases the number of victims, and frequently leads to alcoholism in other family members, invariably leaving deep scars. In this specially-adapted Phil Donahue program, Suzanne Somers tells of growing up in an alcoholic family; she is joined by members of her family, all of whom have struggled with and overcome alcoholism. (28 minutes, color). Films for the Humanities and Sciences, Princeton, NJ.

FEMALE ALCOHOLISM. This program examines the changing stereotype of the female alcoholic and analyzes some case histories of alcoholic women. It explains the dangers of drinking during pregnancy, the effect of fetal alcohol syndrome on newborns, and the emotional effect on children of being raised by an alcoholic mother. (19 minutes, color). Films for the Humanities and Sciences, Princeton, NJ.

— Chapter 8 —
Alcohol: A Behavioral Perspective

Unlike chapter 7, this chapter focuses on alcohol from a behavioral view point. Divided into ten major subheadings, the chapter begins by discussing the pervasive monetary and social costs of alcohol use. Following this, in the next section, the history of alcohol use in America, cultural considerations, and attitudes about drinking are discussed. Next, how widespread is alcohol use and abuse, and how alcohol is used by: the social classes; people of varying religious denominations; different regions of the U.S.; males in comparison to females; different age groups; youth; traditional and nontraditional college student populations; the elderly; different racial and ethnic groups; gay populations; and the homeless. Next, drinking and driving are discussed including a self-check attitudinal test on pages 210-212 of the text that you can administer to students with scoring directions immediately following the attitudinal test on page 212.

Next, alcoholism is defined with a description of whom is likely to be an alcoholic. Next, the explanations of the causes and effects of alcoholism are elaborated on. Two large-scale models, the moral and medical models are discussed and psychological factors -- psychoanalytic theory, psychopathological and social learning and reinforcement theories are explored as explanations of alcoholism. In this section alcoholic types are listed and social and psychological reactions to the alcoholic are presented in seven stages. This section also includes a discussion of what constitutes destructive support, organizations for victims of alcoholics, and concludes with a section on sociological explanations.

The last section in this chapter includes prevention of alcohol abuse followed by another section on the treatment of alcoholism. In this section, rehabilitation methods; aspects of treatment methods; getting through withdrawal; psychological and behavioral therapies; and helping agencies for alcohol rehabilitation concludes this information-rich chapter.

TRUE OR FALSE

1. Alcohol consumption has risen in the United States over the last 10 years. — Ans. F p. 193

2. Ethanol is the psychoactive ingredient that causes alcoholic beverages to be intoxicating. — Ans. T p. 194

3. In the U.S., most people see ethanol as a social substance rather than as the drug that it is. — Ans. T p. 194

4. The temperance movement originally had moderate drinking as its goal and not abstinence. — Ans. T p. 195

5. The temperance movement of the early 1800's was intended for a decline in alcohol consumption.   Ans. T p. 196

6. Throughout history, few countries besides America have passed laws concerning the prohibition of alcohol.   Ans. F p. 196

7. The Twenty First Amendment to the Constitution called for the prohibition of alcohol use in America.   Ans. F p. 196

8. An ambivalent culture is a culture in which alcohol is morally and socially prohibited.   Ans. F p. 197

9. The higher the household income the higher the percentage of those who drink.   Ans. T p. 204

10. According to a 1994 Gallup Poll, approximately 50 percent of men and 45 percent of women drank.   Ans. F p. 204

11. Overall, Northerners drink <u>significantly</u> more than Southerners.   Ans. T p. 204

12. Binge drinking is consuming more than five drinks a week.   Ans. F p. 206

13. Alcohol use tends to decrease as people grow older.   Ans. F p. 207

14. The third leading cause of injury deaths in the United States is motor vehicle crashes.   Ans. F p. 209

15. Most people who consume alcohol develop into problem drinkers.   Ans. F p. 213

16. Problem Drinkers are known as abusers of alcohol, and they are often alcoholics.   Ans. T p. 213

17. Alcoholism is a state of physical and psycological addiction to the psychoactive substance ethanol.   Ans. T p. 214

18. More and more in the United States, alcoholism is being seen and treated as a disease rather than a weakness.   Ans. T p. 215

19. The <u>Moral Model</u> holds the view that excessive alcohol consumption results from a breakdown of social control within a group to which an individual belongs.   Ans. F p. 216

20. According to the Medical Model on Alcoholism, alcohol abuse is a disease that is inate and biologically determined in specific individuals.   Ans. T p. 217

21. In the United States, agreement on the Medical model was so widespread, that in 1970, the U.S. Government passes the Hughes Act, making it the only acceptable model in the field.
Ans. F
p. 217

22. Alcoholism is hard to research because laboratory rats refuse to voluntarily drink alcohol.
Ans. F
p. 217

23. Studies show that 85-90 percent of alcoholics have had a "close alcoholic relative."
Ans. F
p. 218

24. According to Freudian psychoanalytical explanations of addiction to alcohol, deep-seated anxieties that revolve around conflict are the main contributors to addiction.
Ans. T
p. 218

25. According to the social learning theory, alcohol use and later abuse result from early socialization experiences.
Ans. T
p. 219

26. In outlining the social and psychological reactions to the alcoholic, five critical stages were outlined, one of which was the "breakout" stage.
Ans. F
p. 220

27. Enablers are those persons close to the alcholic addict who deny or make "excuses" for excessive drinking.
Ans. T
p. 221

28. Some of the common signs of small children of alcoholics are excessive crying, bed weting, and sleep problems.
Ans. T
p. 221

29. Primary Prevention is a means of techniques that are aimed at more experimental users.
Ans. F
p. 222

30. Most professionals (physicians, social workers, counselors, etc.) feel that Alcoholics Anonymous should be used in support of other forms of therapy, not as a complete form of treatment itself.
Ans. T
p. 226

MULTIPLE CHOICE

1. The term "social substance" refers to when
   a. alcohol is clearly perceived as a drug
   b. alcohol is perceived as a destructive drug
   c. alcohol is not perceived as a drug
   d. combined with other illicit-type drugs
Ans. c
p. 194

2. During the American colonial days, rum _____.
   a. was brought to Africa to pay for slaves
   b. was illegal
   c. could be consumed everyday except on the Sabbath
   d. was non-existent
Ans. a
p. 195

3. An individual's expectation of what a drug will Ans. c
   do to his/her personality is referred to as: p. 196
   a. speakeasy
   b. alcohol determinant
   c. set
   d. setting

4. The Amendment that repealed the Prohibition Ans. d
   Act was: p. 196
   a. Twenty-Second
   b. Eighteenth
   c. Fifteenth
   d. Twenty-First

5. According to David Pittman, which of the following Ans. b
   is not a term used to describe one of the four p. 197
   cultural viewpoints regarding alcohol consumption:
   a. Abstinent Culture
   b. Autonomous Culture
   c. Permissive Culture
   d. Ambivalent Culture
   e. Overpermissive Culture

6. An Ambivalent Culture is one in which: Ans. b
   a. alcohol is morally and socially prohibited p. 197
   b. alcohol is prohibited as well as accepted in
      certain sectors
   c. alcohol consumption is accepted
   d. excessive use of alcohol is permitted

7. In which social class is there a greater likli- Ans. a
   hood of drinking? p. 204
   a. upper class
   b. middle class
   c. lower class
   d. all social classes consume about the
      same

8. Which American minority group has the highest Ans. c
   rate of abstention, the lowest rate of heavy p. 207
   drinking, and the lowest level of drinking-
   related problems?
   a. Native Americans
   b. African-Americans
   c. Asian-Americans
   d. Mexican-Americans

9. Which of the following attributes have been said to be the cause of lower alcohol use in elderly persons:
    a. decreased income
    b. chronic health problems
    c. changes in lifestyle resulting from retirement
    d. the diminished capacity found among peer members
    e. all of the above

    Ans. e
    p. 207

10. According to the Cooperative Commission on the Study of Alcoholism, one of the characteristics of a person likely to develop trouble with alcohol is:
    a. often drinking to a state of intoxication
    b. drinking a beer after work
    c. being a member of a specific culture where alcohol is acceptable
    d. dancing wild after drinking on the weekends

    Ans. a
    p. 216

11. The most effective forms of punishment for alcoholics under the Medical Model is(are):
    a. revocation of driver's license
    b. fines
    c. mandatory alcohol or driving education courses
    d. all of the above
    e. none of the above

    Ans. e
    p. 217

12. One of the approaches to therapy for alcoholics using the Moral Model is
    a. promoting lifestyle changes
    b. increasing fiscal responsibility
    c. emphasising value clarification
    d. support longterm goals of family

    Ans. c
    p. 217

13. Children of alcoholics are _____ times more likely to become alcoholic than children of nonalcoholics.
    a. two
    b. zero
    c. three
    d. four

    Ans. d
    p. 218

14. Clinical psychologists and psychiatrists with a psychoanalytic perspective have described alcoholics in treatment as
    a. healthy
    b. maladjusted
    c. mature
    d. willing to improve their condition

    Ans. b
    p. 218

15. Three of the six types of alcoholics as described by Jellinek are
    a. delta, kappa, sigma
    b. sigma, lambda, gamma
    c. delta, epsilon, zeta
    d. zeta, eta, theta

    Ans. c
    p. 219

16. Alcoholism occurs more frequently in which of the following?
    a. children of alcoholics
    b. adopted children with alcoholic biological parents
    c. cultures which are ambivalent to alcohol
    d. only a and b

    Ans. d
    p. 221

17. Codependency is defined as the behavior displayed by _____, who identify with the alcoholic addict and cover up the excessive drinking behavior.
    a. nonaddicted family members
    b. addicted business associates
    c. addicted family members
    d. local liquor store clerks

    Ans. a
    p. 221

18. The idea of cultural values and norms causing an independant outcome of drinking behavior is called _____.
    a. cultural alcoholism
    b. ethnic comportment
    c. cultural consumption
    d. cultural comportment

    Ans. d
    p. 222

19. About _____ percent of alcoholics are men and women who are married, living with their families and who hold jobs.
    a. 20
    b. 70
    c. 90
    d. 40

    Ans. b
    p. 223

20. Alcoholism is a treatable disease because _____ of those affected can recover.
    a. one-fourth
    b. one-third
    c. one-half
    d. two-thirds

    Ans. d
    p. 223

21. Delirium tremens (DT's) is a condition that may occur _____.
    a. during a highly intoxicated state
    b. during alcohol withdrawal
    c. immediately before a drinker passes out from an alcohol overdose
    d. none of the above

    Ans. b
    p. 224

ESSAYS

1. How costly is alcohol abuse in our society? Substantiate your answer by citing some figures and costs of alcohol consumption as outlined in the first section of chapter 7, Alcohol.

2. Outline some of the main events regarding alcohol use in America. In your discussion include "home brew," some basic history of rum use, whiskey production, drinking patterns during the Jefferson era (1800-1808), temperance movement, and the role of "speakeasies."

3. What are congeners?

4. Describe the short and long term effects of alcohol abuse. In your discussion of *short-term* effects of alcohol, include disinhibition, hangover, and diuretic action of alcohol. In your discussion of *long-term* effects of alcohol include a discussion of how heavy alcohol consumption affects the body's central organs and what is cirrhosis of the liver.

5. What were some of the main findings regarding alcohol use and social class differences, religious differences, regional, gender, and age differences?

6. Why do you think that over 91 percent of high school seniors have tried alcohol and that two-thirds of high school seniors are current drinkers? Cite three to five specific reasons for such widespread use of alcohol by high school students.

7. Begin by first listing the three stages of alcohol addiction outlined by Jellinek then critique the three stages.

8. Of all the prevention programs listed on pages 203-204 which includes: basic prevention methods; specific social environments; individual characteristics; applied prevention; and programs to change individual behaviors; list five factors you think are the most important.

SUPPLEMENTARY MEDIA

CODEPENDENTS. This thirty minute video presents dramatic vignettes played out from the point of view of Nancy Blaine, a codependent and wife of an alcoholic and trying to hold a family together in the face of this debilitating disease. Features two health professionals from Harvard Medical School who analyze Nancy's behavior in each dramatization, showing why alcoholism is a family disease. Available from Indiana University Audio-visual Library, Bloomington, Indiana 47405-5901.

CRUEL SPIRITS, ALCOHOL AND VIOLENCE. This thirty-two minute video documents the relationship between alcohol abuse and violence. Shows footage of community violence, emergency rooms, prisons, treatment centers, and police reports to confirm the connection between alcohol and violent incidents. Explores recent statistics and long term studies, incorporating interviews with medical personnel, researchers, police officials, prisoners, and treatment center residents. Examines alcohol as a factor for both victims and perpetrators of violence. Available from Indiana University Audio-visual Library, Bloomington, Indiana 47405-5901.

FETAL ALCOHOL SYNDROME. This thirteen minute 16mm film reveals the serious mental and physical defects often found in the children of mothers who drank heavily during pregnancy. Shows Kenneth, a 21-month-old boy, and his two school-age sisters who exhibit the defects, such as shortened eye slits, flat facial profile, small brain size, and learning impairments. Presents Dr. David Smith, professor of pediatrics at the University of Seattle, who has characterized the syndrome and has found that the effects caused during prenatal development are irreversible. Discusses rat research which supports Dr. Smith's theory and concludes with an interview with a mother whose drinking has seriously affected her child. Available from Indiana University Audio-visual Library, Bloomington, Indian 47405-5901.

ALCOHOL'S EFFECT ON THE BODY. This program illustrates the wide-ranging negative effects of alcohol on the human body and on society. Dr Ruth Liver delivers the facts on alcohol in the body; drugs and alcohol strain relationships on our soap parody. "The Young and the Breathless": and a fitness instructor describes how alcohol can ruin your appearance. (15 minutes, color) Films for the Humanities and Sciences Princeton, NJ.

ALCOHOL AND THE FAMILY: BREAKING THE CHAIN. This program analyzes the signs of alcoholism and shows how a family member, coworker, or friend can help break the chain; discusses the impact of alcoholism on the children of alcoholics; and evaluates the options and prognosis for alcoholism treatment. (25 minutes, color). Films for the Humanities and Sciences, Princeton, NJ.

CHILDREN OF ALCOHOLICS. Alcohol abuse means physical or psychological abuse of the family as well, which geometrically increases the number of victims, and frequently leads to alcoholism in other family members, invariably leaving deep scars. In this specially-adapted Phil Donahue program, Suzanne Somers tells of growing up in an alcoholic family; she is joined by members of her family, all of whom have struggled with and overcome alcoholism. (28 minutes, color). Films for the Humanities and Sciences, Princeton, NJ.

FEMALE ALCOHOLISM. This program examines the changing stereotype of the female alcoholic and analyzes some case histories of alcoholic women. It explains the dangers of drinking during pregnancy, the effect of fetal alcohol syndrome on newborns, and the emotional effect on children of being raised by an alcoholic mother. (19 minutes, color). Films for the Humanities and Sciences, Princeton, NJ.

− Chapter 9 −
Narcotics (Opioids)

The colorful history of opium and the narcotic drugs are traced from the ancient Egyptians to modern day. Included in this historical overview are discussions of the medicinal uses, social impacts and attitudes, and attempts to regulate these addictive substances. Next, the pharmacological effects and clinical uses of the narcotics are detailed and contrasted with their abuse potential. The major narcotic drugs, such as heroin and morphine, are discussed relative to their abuse and dependence properties, as well as their clinical applications. A brief discussion of miscellaneous drugs used for pain management concludes the chapter and includes agents such anesthetics and sedatives.

TRUE OR FALSE

1. Narcotics mimic the effects of the endorphins. — Ans. T p. 233

2. The lifestyles of "chippers" are dramatically altered because of their dependence on narcotics. — Ans. F p. 233

3. Cocaine is still legally classified as a narcotic. — Ans. T p. 234

4. The word "narcotic" comes from a Greek word that means to excite or activate. — Ans. F p. 234

5. Opium was a great problem for several centuries in China. — Ans. T p. 236-237

6. Smoking narcotics has only been popular since the 1980's. — Ans. F p. 236

7. Hypodermic needles were used extensively during the Civil War to administer morphine for pain and dysentary. — Ans. T p. 237

8. Opiate addiction was not a significant problem in the U.S. until after 1950. — Ans. F p. 237

9. Narcotics are the most potent of the commonly used analgesics. — Ans. T p. 239

10. It is likely that the pain relieving effects of acupuncture is due to the release of endorphins. — Ans. T p. 239

11. Narcotic analgesics alter the perception of pain. — Ans. F p. 239

12. Because of their abuse potential, narcotic drugs are no longer used very much for therapy. Ans. F p. 239

13. Narcotics are often combined with aspirin to treat pain. Ans. T p. 239

14. Because of concern about dependence, doctors often do not prescribe enough narcotics to often control patients' pain. Ans. T p. 240

15. The most common side effect of using narcotics to relieve pain is constipation. Ans. T p. 240

16. Narcotics work by blocking the opioid receptors. Ans. F p. 240

17. Narcotic analgesics can depress respiration. Ans. T p. 240

18. A narcotic analgesic would most likely be used to treat severe diarrhea. Ans. T p. 240

19. Most pains can be relieved by narcotic analgesics. Ans. T p. 240

20. Heroin is the most likely of the opioid narcotics to be severely abused. Ans. T p. 241-242

21. Tolerance to narcotics doesn't occur until after several weeks of use. Ans. F p. 241

22. Tolerance to one narcotic means tolerance to other narcotics has also occurred. Ans. T p. 241

23. Heroin is sometimes cut (combined) with fentanyl. Ans. T p. 242

24. Alcohol use decreases the chances of heroin causing dangerous effects. Ans. F p. 242

25. The abuse of narcotics is only an inner city problem. Ans. F p. 244

26. The use of heroin has declined since 1987. Ans. T p. 244

27. Most people are "hooked" on heroin after the first administration. Ans. F p. 245

28. Because of the publicized dangers concerning cocaine use, some dealers of this stimulant have switched to selling heroin. Ans. T p. 245

29. Withdrawals from chronic heroin use often are lethal if not treated. Ans. F p. 246

30. Almost 50% of the heroin addicts in this country have been exposed to the AIDS virus. — Ans. T p. 246

31. When "the monkey is on his back," a heroin user must take more narcotics to avoid withdrawal symptoms. — Ans. T p. 246

32. Needle sharing by heroin addicts in "shooting galleries" substantially increases the likelihood of getting HIV infection. — Ans. T p. 246

33. The severity of withdrawal from opioids depends on the personality of the user. — Ans. T p. 247

34. Methadone is longer acting than heroin. — Ans. T p. 248

35. Codeine is frequently used to help narcotic addicts stop using heroin or another more addictive drug. — Ans. F p. 248-249

36. The first exposure to narcotics such as morphine may be unpleasant. — Ans. T p. 250

37. All of the narcotics addicts who participate in methadone maintenance programs are eventually withdrawn from the methadone. — Ans. F p. 249

38. Methadone is never abused. — Ans. F p. 251

39. Methadone is shorter acting than heroin. — Ans. F p. 251

40. Fentanyl can be made into "designer" drugs. — Ans. T p. 252

41. MPTP use causes Parkinson's disease. — Ans. T p. 252

42. Codeine is not as effective an analgesic as morphine. — Ans. T p. 253

43. Codeine in cough medicine can be sold without a prescription in some states. — Ans. T p. 253

44. Naloxone is used as an antidote for narcotic overdoses. — Ans. T p. 254

45. When administered alone, naloxone is a potent stimulant. — Ans. F p. 254

MULTIPLE CHOICE

1. Which of the following was the reason for the "opium wars" between China and Great Britain?
   a. The Chinese attacked the British because of their mistreatment of Chinese citizens.
   b. The British attacked the Chinese because of Chinese opposition to British opium trade in China.
   c. The British attacked the Chinese because the Chinese were shipping opium to England.
   d. The Chinese attacked the British because they would not allow the Chinese to ship opium to England.

   Ans. b
   p. 236

2. Select the <u>incorrect</u> statement.
   a. The invention of the hypodermic needle in 1853 worsened the problem of morphine addiction.
   b. Morphine was injected extensively during the Civil War causing a high rate of morphine addiction in the soldiers returning home after the war.
   c. Patent medicines containing opium did not significantly contribute to opium addiction in this country at the turn of the century.
   d. The average opiate addict in the early 1900s was a middle aged, southern, white woman.

   Ans. c
   p. 237

3. Which of the following first extracted and purified the active ingredients in opium?
   a. Sigmund Freud
   b. Frederick Serturner
   c. Alexander Wood
   d. Christopher Wren

   Ans. b
   p. 237

4. Which of the following did not influence the opiate problem around 1900?
   a. Chinese laborers
   b. bored housewives
   c. hypodermic syringes
   d. patent medicines

   Ans. b
   p. 237

5. Which of the following was the first use of heroin in the United States?
   a. sleepaid
   b. antidepressant
   c. Analgesic
   d. Cough suppressant

   Ans. d
   p. 238

6. In 1914, the regulation of opium began with the _____. Ans. a
    a. Harrison Narcotic Act   p. 238
    b. Tyler Drug Act
    c. McKinley Drug Control Act
    d. none of the above

7. Which of the following is not a clinical use of narcotics? Ans. b
    a. analgesic   p. 239
    b. reduce blood pressure
    c. treat diarrhea
    d. antitussive

8. Which of the following statements is <u>incorrect</u> concerning the effect of narcotics on pain? Ans. c
    a. They relieve pain because they stimulate opioid receptors in the brain and spinal cord.   p. 240
    b. Often narcotic use doesn't block the pain, but makes the painful experiences more tolerable.
    c. Narcotics usually cause diarrhea.
    d. High doses of narcotics will suppress respiration.

9. Codeine is a(n) _____. Ans. b
    a. antipyretic   p. 240
    b. antitussive
    c. NSAID
    d. both a and b

10. Which of the following are possible side effects of narcotic analgesics? Ans. d
    a. respiratory depression   p. 240-241
    b. constipation
    c. decreased blood pressure
    d. all of the above

11. Heroin is classified as a _____ drug, and cannot be used clinically in the United States. Ans. a
    a. Schedule I   p. 241
    b. Schedule II
    c. Schedule III
    d Schedule IV

12. Which of the following is a clinical use for the drug naloxone? Ans. a
    a. to treat the person who has overdosed on narcotics   p. 241
    b. to relieve withdrawals in the narcotic abuser who has become dependent
    c. to treat severe pain
    d. to relax a patient who is very stressed

13. Select the incorrect statement.　　　　　　　　　　Ans. c
    a. Heroin was originally marketed as a cough　　　p. 241-
       suppressant.　　　　　　　　　　　　　　　　　　　242
    b. Heroin was originally developed in order to
       replace morphine with a less addictive
       narcotic.
    c. Heroin is currently classified as a Schedule
       II drug in this country.
    d. When injected, heroin tends to be more addict-
       ing than most other narcotics because of a
       rapid intense feeling of euphoria.

14. Heroin that is sold to users in the United States　Ans. b
    is usually only _____ pure.　　　　　　　　　　　　p. 242
    a. 30%-50%
    b. 3%-5%
    c. 50%-80%
    d. 5%-8%

15. Which of the following drugs is most commonly　　　Ans. a
    combined with heroin?　　　　　　　　　　　　　　　　p. 242
    a. Alcohol
    b. Cocaine
    c. LSD
    d. Valium

16. Which of the following is called "speedballing?"　Ans. b
    a. Combining heroin with alcohol　　　　　　　　　p. 242
    b. Combining heroin with cocaine
    c. Combining heroin with opium
    d. Combining heroin with caffeine

17. Select the correct statement.　　　　　　　　　　　Ans. d
    a. Pure heroin is a brown powder.　　　　　　　　　p. 242-
    b. Heroin sold on the street usually has a　　　　243
       purity greater than 50%.
    c. Heroin always causes antisocial behavior in
       its users.
    d. Surveys have shown that from 1987 to 1991
       heroin use has declined in the U.S.

18. Select the incorrect statement.　　　　　　　　　　Ans. d
    a. Most heroin addicts "mainline" the drug.　　　 p. 242-
    b. Withdrawal symptoms from heroin use resemble　247
       a bad cold.
    c. It requires approximately 5-7 days to recover
       from the physical symptoms of heroin
       withdrawal.
    d. Withdrawal symptoms from each narcotic are
       very different.

19. Mainlining refers to _____.
    a. injecting drugs under the skin
    b. intramuscular injection
    c. sniffing the powder drug
    d. intravenous injection

    Ans. d
    p. 246

20. Which of the following are symptoms of withdrawal from heroin?
    a. runny nose
    b. constipation
    c. vomiting
    d. both a and b

    Ans. d
    p. 246-247

21. How long does it take for withdrawals to begin once an addict stops using heroin?
    a. 1 hours
    b. 4-6 hours
    c. 1-2 days
    d. 1 week

    Ans. b
    p. 246-247

22. Select the <u>incorrect</u> statement.
    a. Morphine is a more potent analgesic than heroin.
    b. Narcotic use often causes nausea and constricted pupils.
    c. Morphine is classified as a Schedule II controlled substance.
    d. Tolerance to the effects of morphine can develop quickly if used continually.

    Ans. a
    p. 250

23. Select the <u>correct</u> statement concerning methadone.
    a. Methadone is as addictive as heroin if injected.
    b. It does not have cross tolerance with other narcotic drugs.
    c. Post addiction craving for heroin is not suppressed by methadone use.
    d. Methadone does not have the same side effects as heroin.

    Ans. a
    p. 250-251

24. Which of the following statements concerning narcotic "chippers" is <u>incorrect</u>?
    a. They are usually very open about their problem.
    b. They often are not accepted into methadone maintenance programs.
    c. They do not use heroin daily.
    d. They are usually meticulous and in control with their narcotic use.

    Ans. a
    p. 251

25. Methadone is effective for about _____. Ans. c
    a. 1-2 hours p. 251
    b. 10-12 hours
    c. 24-36 hours
    d. 36-48 hours

26. Select the correct statement concerning fentanyl. Ans. c
    a. It is not approved for any clinical use. p. 251-252
    b. It is less potent than heroin.
    c. It can be chemically modified and still retain its addictive properties.
    d. It is easy to distinguish street heroin and fentanyl.

27. Which of the following is the generic name for Demerol? Ans. b
    a. Codeine p. 252
    b. Meperidine
    c. Hydromorphone
    d. Pentazocine

28. Talwin is the trade name for _____. Ans. c
    a. codeine p. 253
    b. meperidine
    c. pentazocine
    d. propoxyphene

29. Codeine is most likely to be used to relieve _____. Ans. d
    a. muscle cramps p. 253
    b. constipation
    c. headaches
    d. coughing

30. Which of the following best describes "LAAM?" Ans. b
    a. An antagonist used to treat narcotic overdoses p. 248
    b. A long acting narcotic to prevent narcotic withdrawal
    c. A very potent heroin-like drug of abuse
    d. An aspirin-like drug combined with narcotics to improve their analgesic action

31. Which of the following is the most common clinical use of narcotics today? Ans. a
    a. relieve pain p. 239-240
    b. relieve coughing
    c. relieve diarrhea
    d. to treat cocaine dependence

32. Which of the following transmitters likely contributes to narcotic dependence?
    a. acetylcholine
    b. GABA
    c. noradrenaline
    d. dopamine

    Ans. d
    p. 240

33. Which of the following is not likely to be a significant side effect of narcotics?
    a. diarrhea
    b. slow breathing
    c. drowsiness
    d. nausea

    Ans. a
    p. 240-241

34. Heroin is classified in which of the following Schedules by the DEA?
    a. I
    b. II
    c. III
    d. IV

    Ans. a
    p. 241

35. Select the <u>incorrect</u> statement concerning the narcotic, fentanyl.
    a. it is less potent than morphine
    b. it is used in transdermal patches
    c. it has been used to create "designer" narcotics
    d. it has been used to substitute for heroin

    Ans. a
    p. 251-252

36. Which of the following narcotics is most like "MPTP?"
    a. codeine
    b. heroin
    c. morphine
    d. meperidine

    Ans. d
    p. 252

37. Use of MPTP causes which of the following diseases?
    a. Alzheimer's disease
    b. cancer
    c. Parkinson's disease
    d. epilepsy

    Ans. c
    p. 252

38. Which of the following is the principal therapeutic use of dextromethorphan?
    a. relieve pain
    b. treat diarrhea
    c. treat narcotic addiction
    d. relieve coughing

    Ans. d
    p. 253-254

ESSAYS

1. Outline the stages of heroin dependence.

2. What advantage is there in substituting methadone for heroin

in an addict who is not interested in breaking his or her narcotic dependence?

3. Describe the factors that contributed to the problems of narcotic addiction in the U.S. around 1900.

4. Describe the symptoms associated with narcotic withdrawal.

5. What have been good and bad consequences of the synthesis of MPTP?

6. How might naloxone be used to treat narcotic dependence?

SUPPLEMENTARY MEDIA

NARCOTIC FILE: THE SOURCE. This twenty-eight minute film discusses where narcotics originate. Legal efforts to reduce drug traffic are reviewed. Available from Indiana University Audio-visual Library, Bloomington, Indiana 47405-5901.

NARCOTICS-WHY NOT? This fifteen minute film presents a series of extemporaneous interviews with teenagers and young adults who have been narcotic addicts. It shows ends to which addicts might go to obtain drugs. It interviews two addicts who relate the misery of being an addict. Available from Indiana University Audio-visual Library, Bloomington, Indiana 47405-5901.

THE OPIUM WARLORDS. This film is an extraordinary documentary, shot during 18 months in the guerrilla-held, opium producing regions of Burma, is the first inside report ever made of the Southeast Asia narcotics business. A third of the world's black-market heroin comes from the Shan states of Burma, where rival armies betray and ambush each other's convoys in a desperate struggle for control of the opium business. This film details the circumstances surrounding the Shan army's proposal to sell the U.S. government 400 tons of opium a year. Available from Kent State Film Library, Kent, Ohio 44242.

METHADONE: WHERE ARE WE? This twenty minute video examines the use and effectiveness of methadone as a treatment for narcotic dependence; the biological effects of methadone; and societal attitudes and stigmas regarding methadone treatment and patients. Available from National Clearinghouse for Alcohol and Drug Information (NCADI) P.O. Box 2345, Rockville, MD 20847-2345 or call (800) 729-6686.

HOOKED ON HEROIN: FROM HOLLYWOOD TO MAIN STREET (GX5149-VHS). This fifty-two minute video interviews heroin addicts from diverse backgrounds and discusses why heroin addiction is again becoming a major drug abuse problem. Available from "Films for the Humanities and Sciences," P.O. Box 2053, Princeton, NJ 08543 or call (800) 257-5126.

Chapter 10

Stimulants

Stimulants are substances that increase the activity of the central nervous system (CNS). The user often experiences an initial sense of energy and euphoria which accounts for the addictive properties of these drugs. Chapter 10 first deals with the major stimulants such as amphetamines and cocaine. The history of these substances is presented, followed by an explanation of how they alter the activity of the brain, their current therapeutic uses and side effects. These sections include a detailed discussion of the patterns of misuse and abuse for amphetamines and cocaine and the principal approaches used to treat dependence on the major stimulants. Included are sections that specifically deal with use of the major stimulants by athletes and the effect of cocaine on pregnancy.

The chapter concludes with a detailed discussion of minor CNS stimulants, in particular caffeine. In this section the chemical properties of caffeine are described and the major sources of caffeine are discussed. Some interesting history of the origins of caffeine-containing beverages is included. The effects of caffeine on the CNS and cardiovascular systems are presented. Details on abuse potential and dependence conclude the section on caffeine. Chapter 10 concludes with a short discussion of OTC sympathomimetics as minor CNS stimulants. The abuse of these drugs in "look-alike" or "act-alike" nonprescription products is explained.

TRUE OR FALSE

1. The major stimulants are called "downers" since they cause a rebound depression.
   Ans. F
   p. 260

2. Amphetamine was available in OTC nasal decongestant inhalers until the 1960-70s.
   Ans. T
   p. 261

3. At one time amphetamines were promoted as a treatment for bedwetting, depression, hiccups, and schizophrenia.
   Ans. T
   p. 261

4. Amphetamines were originally used in nasal inhalers.
   Ans. T
   p. 261

5. Amphetamines are commonly called jolly beans, speed, bennies, uppers, footballs, and hearts.
   Ans. T
   p. 262

6. Persistently singing one note, repeating a phrase of music, or repeatedly cleaning the the same object are examples of behavioral stereotype that can be caused by heavy amphetamine use.
   Ans. T
   p. 262

7. Chronic high doses of amphetamines have not been shown to permanently damage the brain.
   Ans. F
   p. 262

8. Ritalin is a form of cocaine used to treat attention deficit children.
   Ans. F
   p. 263

9. The most common FDA-approved use of amphetamines, today, is as a diet aid to treat obesity.
   Ans. T
   p. 263

10. Tolerance does <u>not</u> occur to the anorexiant effects of the amphetamines.
    Ans. F
    p. 263

11. In the last 10 years, the prescription use of amphetamines has increased in the country.
    Ans. F
    p. 264

12. Amphetamine drugs are relatively easy to synthesize if the chemicals are available.
    Ans. T
    p. 264

13. The amphetamines can cause seizures when high doses are administered.
    Ans. T
    p. 265

14. Nonhealing ulcers, considerable weight loss, kidney damage, and seizures are all common results of continued use of massive doses of amphetamines.
    Ans. T
    p. 265

15. Speed freaks often live and use drugs together with addicts who use LSD.
    Ans. F
    p. 266

16. Amphetamines and cocaine are often used together with barbiturates and benzodiazepine.
    Ans. T
    p. 266

17. Some psychiatrists have used MDMA (ecstasy) in psychoanalysis.
    Ans. T
    p. 268

18. In the late 1800s the use of cocaine in a popular wine was endorsed by the Catholic Pope.
    Ans. T
    p. 271

19. Throughout the 20th century the dangerous nature of cocaine use has been recognized by the general public.
    Ans. F
    p. 271-274

20. Hundreds of years ago, coca was used in South America to pay the slaves as well as pay taxes to the Catholic Church.
    Ans. T
    p. 271

21. From the start, Sigmund Freud warned about the dangers of cocaine.
    Ans. F
    p. 272

22. During the third cocaine era, cocaine was only a problem of the rich and famous.
    Ans. F
    p. 273-274

23. Use of crack cocaine tends to be more dangerous than cocaine snorting.
Ans. T
p. 277-278

24. "Freebasing" is converting cocaine to an alkaline form so that it can be smoked.
Ans. T
p. 277

25. "Crack" is a form of alkaline (basic) cocaine.
Ans. T
p. 278

26. Amphetamine and cocaine have similar pharmacological effects.
Ans. T
p. 280

27. The CNS effects of cocaine have a slower onset but result in a more intense high than those of amphetamines.
Ans. F
p. 280

28. Cocaine is still used clinically as a local anesthetic.
Ans. T
p. 281

29. Cocaine use does not cause physical dependence.
Ans. F
p. 281

30. Physical withdrawal from cocaine dependence is usually as intense as that associated with CNS depressant dependence.
Ans. F
p. 281

31. Cocaine was the first local anesthetic used routinely in modern-day medicine.
Ans. T
p. 281

32. The first local anesthetic used for modern day surgery was cocaine.
Ans. T
p. 281

33. The "crash," withdrawal, and extinction are the three main stages of cocaine withdrawal.
Ans. T
p. 282

34. Most cocaine-dependent persons use little alcohol.
Ans. F
p. 283

35. Stimulants are used to relieve anxiety during the "crash" phase of cocaine treatment.
Ans. F
p. 283 (table)

36. Pepsi Cola contains more caffeine than a cup of coffee.
Ans. F
p. 285 (table)

37. A chocolate drink, such as cocoa, contains as much caffeine as does a cup of coffee.
Ans. F
p. 287

38. Caffeine is the principal xanthine in chocolate.
Ans. F
p. 288

39. Caffeine appears to have some analgesic effects.
Ans. T
p. 289

40. Adverse effects usually don't begin until caffeine doses greater than 2gm are consumed.
Ans. F
p. 290

41. Some researchers suggest that consumption of large amounts of caffeine can result in cancers of the bladder, ovaries, colon, and kidneys.
Ans. T
p. 291

42. The absorption rate for caffeine from the gut is highly variable between users.
Ans. T
p. 292

43. Look-alike or act-alike drugs contain illegal stimulants.
Ans. F
p. 292

MULTIPLE CHOICE

1. Amphetamine and cocaine are classified in which of the following Schedules?
    a. I
    b. II
    c. III
    d. IV

    Ans. b
    p. 260

2. Which of the following was the first clinical use for amphetamines?
    a. to treat obesity
    b. to treat insomnia
    c. to treat attention deficit children
    d. to relieve nasal congestion

    Ans. d
    p. 261

3. Anorexiants are drugs that _____.
    a. suppress appetite
    b. treat bedwetting
    c. counteract depression
    d. none of the above

    Ans. a
    p. 261

4. Select the incorrect statement concerning amphetamines.
    a. They were used by soldiers during World War II and the Korean War to counter fatigue.
    b. Truck drivers making long hauls were one of the earliest distribution systems for illicit amphetamines.
    c. Large scale epidemics of amphetamine abuse have only occurred in this country.
    d. Speed is a common slang term for a type of amphetamine.

    Ans. c
    p. 261

5. Which of the following <u>best</u> explains how the amphetamines cause their pharmacological effects?
   a. They cause release of the neurotransmitter, dopamine, norepinephrine and serotonin in the brain.
   b. They directly stimulate the synthesis of dopamine in the brain.
   c. They stimulate the receptors for dopamine and norepinephrine in the brain.
   d. Their mechanism of action is unknown.

   Ans. a
   p. 262

6. Behavioral stereotypy caused by high doses of amphetamines and cocaine is <u>most likely</u> caused by activation of which neurotransmitter?
   a. noradrenaline
   b. GABA
   c. dopamine
   d. acetylcholine

   Ans. c
   p. 262

7. The most common legal use of amphetamines is for the treatment of _____.
   a. narcolepsy
   b. hyperkinesis
   c. obesity
   d. antidepression

   Ans. c
   p. 263

8. Which of the following is <u>not</u> an effect of the amphetamines?
   a. has a calming effect on children with attention deficit disorders
   b. has anorexiant effects
   c. decreases heart rate
   d. elevates blood pressure

   Ans. c
   p. 263-264

9. The term "speed" refers to which of the following?
   a. another term for ecstacy
   b. cocaine
   c. ritalin used illegally
   d. illegal methamphetamine

   Ans. d
   p. 264

10. Which of the following describes "ice?"
    a. cocaine that is smoked
    b. methamphetamine that is smoked
    c. cocaine that is injected
    d. methamphetamine that is injected

    Ans. b
    p. 264

11. Which of the following is <u>least likely</u> to be caused by chronic high doses of amphetamines?
    a. hallucinations
    b. stereotypic behavior
    c. feelings of bugs crawling under the skin
    d. extreme hunger

    Ans. d
    p. 264-265

12. Which of the following is least likely to be a property of designer drugs?
    a. chemically related to common drugs of abuse
    b. when they first appear, they may not be scheduled by the DEA
    c. often skillfully marketed on the "street" under attractive and exotic names
    d. drugs which have not previously been synthesized

    Ans. d
    p. 266

13. Which of the following best describes the effects of ecstacy?
    a. a CNS depressant
    b. a pure hallucinogen
    c. a pure stimulant
    d. a drug with stimulant and hallucinogen effects

    Ans. d
    p. 267

14. Ecstasy is classified in which of the following Schedules?
    a. I
    b. II
    c. III
    d. IV

    Ans. a
    p. 267

15. Some designer amphetamines are unique in that they cause both _____ and _____.
    a. excitation; depression
    b. depression; hallucinations
    c. excitation; hallucinations
    d. none of the above

    Ans. c
    p. 268

16. Select the incorrect statement concerning methylphenidate.
    a. mild to moderate CNS stimulant
    b. generic name for Ritalin
    c. commonly used by psychiatrists to treat severe depression
    d. frequently used to treat attention deficit disorder

    Ans. c
    p. 269

17. The amphetamine, Ritalin, is used to treat _____.
    a. fatigue
    b. hyperactive children
    c. menstrual cramps
    d. both a and c

    Ans. b
    p. 269

18. Studies have shown that the first cocaine era began in _____ around 2500 B.C.
    a. North America
    b. South America
    c. Europe
    d. Africa

    Ans. b
    p. 270

19. Select the <u>incorrect</u> statement. Ans. b
    a. Europeans became aware of cocaine during p. 270-
       the second major cocaine era. 272
    b. Coca-Cola was originally a patent medicine
       that contained heroin.
    c. At one time Sigmund Freud advocated the use
       of cocaine to treat drug addiction.
    d. Cocaine was used as a source of energy by
       the ancient South American Indians who lived
       in the regions of the Andean Mountains.

20. In the late 1800s, a wine containing cocaine was Ans. d
    endorsed by _____. p. 271-
    a. the President of the United States 272
    b. the Pope
    c. world royalty
    d. all of the above

21. Select the <u>incorrect</u> statement. Ans. a
    a. In the early 1980s cocaine was viewed by p. 273-
       most people as a very dangerous stimulant. 275
    b. Cocaine is not easily synthesized, so it
       is derived entirely from the coca plant of
       South America.
    c. The coca crop is the most profitable
       crop in Bolivia and Columbia.
    d. Most cocaine sold on the streets has been
       "cut" with adulterants.

22. The _____ of 1914 outlawed the uncontrolled Ans. d
    use of cocaine and coca. p. 273
    a. Brigitte-Sinclair Act
    b. Parkshed Act
    c. Tinley Amendment
    d. Harrison Act

23. In Latin America, coca is traditionally seen as Ans. c
    _____. p. 274
    a. a desirable substance
    b. a big cash crop
    c. both a and b
    d. none of the above

24. The purity of adulterated (or "cut") cocaine Ans. c
    usually ranges from _____. p. 275
    a. 1%-5%
    b. 5%-10%
    c. 10%-60%
    d. 60%-90%

25. Cocaine abuse has _____ since the mid-1980s. Ans. b
    a. increased somewhat p. 275
    b. significantly declined
    c. remained the same
    d. increased dramatically

26. _____ of cocaine produces the most potent effects.
    a. Intravenous administration
    b. "Snorting"
    c. Oral administration
    d. None of the above

    Ans. a
    p. 276-277

27. Which of the following best describes "freebased" cocaine?
    a. cocaine converted to its acidic form for snorting
    b. cocaine converted to its alkaline form for I.V. injection
    c. cocaine converted to its acidic form for smoking
    d. cocaine converted to its alkaline form for smoking

    Ans. d
    p. 277-278

28. Which of the following forms of cocaine administration is the most dangerous and most likely to cause dependence?
    a. oral
    b. smoking
    c. I.V.
    d. snorting

    Ans. b
    p. 278

29. Which of the following statements concerning "crack" is incorrect?
    a. The use of crack continues to increase substantially each year.
    b. Crack is smoked.
    c. The "rush" resulting from using crack is likely due to the release of dopamine.
    d. Crack is relatively inexpensive and readily available.

    Ans. a
    p. 278

30. Select the incorrect statement.
    a. People experiencing withdrawal from cocaine dependence are usually depressed and have a high suicide rate.
    b. Withdrawal from cocaine interferes with one's ability to experience pleasure.
    c. Most cocaine-dependent persons only abuse cocaine.
    d. Cocaine may adversely affect the fetus when used during pregnancy.

    Ans. c
    p. 279-283

31. The mechanism of cocaine's pharmacological action on the brain most closely resembles which of the following groups of drugs?
    a. antipsychotics
    b. antidepressants
    c. antianxiety
    d. pain relievers

    Ans. b
    p. 279

32. Which of the following drugs was used as the first local anesthetic for surgery by modern man?
    a. cocaine
    b. amphetamine
    c. methamphetamine
    d. retalin

    Ans. a
    p. 281

33. The intensity of cocaine withdrawal is proportional to _____.
    a. the amount used
    b. the duration of use
    c. the intensity of use
    d. all of the above

    Ans. d
    p. 281

34. Select the <u>incorrect</u> statement concerning "cocaethylene":
    a. it enhances cardiovascular toxicity of cocaine
    b. it enhances the euphoria of cocaine
    c. it is a "designer" cocaine sold in the streets
    d. it is often found in the body of victims who overdosed on cocaine

    Ans. c
    p. 284

35. Which of the following is the most frequently used CNS stimulant in the world?
    a. caffeine
    b. cocaine
    c. amphetamines
    d. ecstasy

    Ans. a
    p. 285

36. Which of the following drugs is <u>not</u> classified as a xanthine?
    a. caffeine
    b. phenylpropanolamine
    c. theophylline
    d. theobromine

    Ans. b
    p. 285

37. Select the <u>incorrect</u> statement concerning coffee.
    a. It has been used as a medicine by some societies.
    b. Coffee beans are primarily cultivated in South America and East Africa.
    c. Coffee replaced tea in the American Colonies during the Revolutionary War.
    d. The daily use of coffee continues to increase every year.

    Ans. d
    p. 285-286

38. The xanthines are _____.
    a. depressants
    b. antidepressants
    c. stimulants
    d. tranquilizers

    Ans. c
    p. 285

39. Coffee was used initially in Europe for its _____.
    a. euphoriant effects
    b. tranquilizing effects
    c. antidepressants
    d. medicinal value

    Ans. d
    p. 286

40. The most common source of xanthine consumption is through _____.
    a. organic foods
    b. caffeinated beverages
    c. chocolate
    d. decaffeinated beverages

    Ans. b
    p. 286, 287

41. Which of the following is <u>least likely</u> to be caused by the consumption of caffeine?
    a. increases a sense of alertness
    b. increases intelligence
    c. diminishes the sense of boredom
    d. effects are most pronounced on unstimulated, drowsy consumers

    Ans. b
    p. 290

42. Select the <u>incorrect</u> statement.
    a. The symptoms of caffeinism include agitation, tremors, and insomnia.
    b. Caffeine has been proven to cause cancers of the bladder, colon and kidneys.
    c. Psychological dependence on caffeine is usually minor.
    d. Sympathomimetics in OTC cold products can cause jitters and insomnia like caffeine.

    Ans. b
    p. 290-292

43. Which of the following xanthines has the most potent effect on the central nervous system (CNS)?
    a. theobromine
    b. theophylline
    c. caffeine
    d. they are all the same

    Ans. c
    p. 290

44. Which of the following may be the effects of caffeine withdrawal?
    a. headaches
    b. nausea
    c. muscle pain
    d. all of the above

    Ans. d
    p. 291

45. Pseudoephedrine is a _____.
    a. decongestant
    b. diet aid
    c. upper
    d. depressant

    Ans. a
    p. 292

46. Amphetamines _____. Ans. a
    a. have high abuse potential p. 293
    b. are illegal in all cases
    c. are rarely abused
    d. cause decreased blood pressure

47. The effects of caffeine are most pronounced Ans. c
    in a person who is _____. p. 294
    a. susceptible to mood swings
    b. wide awake
    c. tired
    d. none of the above

ESSAYS

1. Give a brief but succinct history of amphetamines followed by their early use and currently approved use.

2. In one or more paragraphs explain each of the three cocaine era's. In your answer, identify major events and opinions of each era.

3. What are the effects of cocaine in comparison to amphetamines?

4. What is crack? How is it made? What makes it unique from other forms of cocaine?

5. What are the advantages and disadvantages of legalizing cocaine? How do you feel about the two alternatives?

6. Select two of the minor stimulants, give a brief history of the two drugs and discuss them in terms of their effect on the body.

7. Explain the following in three to five sentences: uppers; speed; ice; run; speedball; freebasing; and caffeinism.

8. Explain the therapeutic drug treatment used to treat drug dependence.

SUPPLEMENTARY MEDIA

COCAINE PAIN. This twenty-five minute video presents the stories of five cocaine abusers trying to kick the habit, focusing on one particular cocaine rehabilitation program called Cokenders which is designed to help cocaine abusers understand why they have become hooked. Explains the similarities and differences in cocaine and alcohol abuse and looks at the effects of cocaine abuse on health, personal relationships, and mental capacity. Illustrates the fact that the psychological addiction of cocaine is a life-long battle. Available from Indiana University Audio-visual Library, Bloomington, Indiana 47405-5901.

CRACK: DEAD AT SEVENTEEN. This thirteen minute video dramatizes the story of a teenage boy who dies after using crack. Uses slow-motion and dream-like sequences as the boy narrates to explain the events leading up to his death and his thoughts after he dies. Shows how he is now just a statistic, a lesson for other kids to learn, and looks at the pain and suffering he has caused his family. Available from Indiana University Audio-visual Library, Bloomington, Indiana 47405-5901.

CRACK. This twenty eight minute video explores the reasons why people start using crack and effects it has on them. Uses experiences of both users and those who counsel and treat them to explain the warning signs of a potential user, the kinds of behavior that results from long-term use, and how users can be induced to seek treatment. Includes appearances by Dr. Arnold Washton, Director of the National Cocaine Hotline and Bob Stuttman of the U.S. Drug Enforcement Agency. Hosted by Phil Donahue. Available from Indiana University Audio-visual Library, Bloomington, Indiana 47405-5901.

— Chapter 11 —
Tobacco

This chapter begins by introducing the scope of the problem and current use statistics in the use of tobacco. Proceeding this section, the history of tobacco use is discussed with sections on the popularity of tobacco use in the western world, tobacco use in America, tobacco production, and government regulations. Next, the pharmacology of nicotine, namely, the properties of this drug are discussed. Following this, in the next section, the physiological effects are assessed in terms of: mortality rates; chronic illnesses; cardiovascular disease; cancer; broncho-pulmonary disease; and effects on the fetus. This section ends with a discussion on what are clove cigarettes and their effects on health.

Next, smokeless tobacco, another type of tobacco use chewing tobacco and snuff is discussed followed by secondhand and "sidestream" smoke. The next section elaborates who is most likely to smoke and why. In this section, the characteristics of smokers, the reasons usually given for why someone would smoke, and youth and smoking are discussed. The next section focuses on how to stop smoking. Here, withdrawal and readdiction, successful behavioral treatments, and stop smoking aids such as nicorette gum and the nicotine patch are discussed.

The last section, looks at the economic incentives of the tobacco industry and why and how this industry promotes tobacco (largely cigarette) use. Included in this final section is a discussion of how tobacco is a "gateway" drug, how smoking in public is currently being curtailed, how smokers are responding to increasingly restrictive public banning of smoking, and a final profile of what severely addicted and/or least likely to quit smokers share in common.

TRUE OR FALSE

1. Tobacco is the second largest cash crop in the United States.
Ans. F
p. 299

2. Nicotine is believed to be as addictive as cocaine and heroine.
Ans. T
p. 299

3. Uninhailed smoke from a lit cigarette has higher concentrations of carbon monoxide, nicotine, and ammonia.
Ans. T
p. 299

4. In the 1600s, it was not uncommon for someone to be put to death for smoking.
Ans. T
p. 301

5. Cigarette smoking has increased in the United States but declined in Third World countries.
Ans. F
p. 306

6. Nicotine is highly toxic and has been used as an insecticide. Ans. T p. 306

7. Nicotine first depresses and then stimulates the nervous system. Ans. F p. 306

8. Lung cancer is the leading cause of cancer death in the United States. Ans. T p. 307

9. Persons with a low amount of an enzyme called alpha-1-antitrypsin are less likely to develop emphysema. Ans. F p. 309

10. Babies born to mothers who smoke have a lower average body weight and length and yet have a larger head circumference. Ans. F p. 309

11. The below-average weight of babies born to smokers is caused by carbon monoxide and nicotine. Ans. T p. 309

12. Sudden infant death syndrome (SID) is the unexpected and unexplainable death that occurs while infants are sleeping. Ans. T p. 309

13. Chewing and snuff tobacco are two types of smokeless tobacco. Ans. T p. 310

14. Rapid smoking involves inhaling a cigarette every six seconds until the smoker can no longer bear smoking. Ans. T p. 319

15. Snuff consumption has increased in the early eighties. Ans. T p. 310

16. Women are _more likely_ to smoke cigarettes than men. Ans. F p. 302

17. The more education you have, the _less likely_ you will smoke. Ans. T p. 315

18. Consistent and related behaviors that occur together are called patterns of behavior. Ans. T p. 322

MULTIPLE CHOICE

1. The tobacco primary plant species cultivated on the American continent is called _____. Ans. d p. 302
   a. Nicotina foliage
   b. Nicotina cerebellum
   c. Tobacco topina
   d. Nicotine tabacum

2. Which two states are the leading producers of tobacco in the United States?
   a. North Carolina and Kentucky
   b. California and Idaho
   c. Alabama and South Carolina
   d. Virginia and South Carolina

   Ans. a
   p. 302

3. What is the total amount spent by the American consumer for tobacco products a year?
   a. $15 billion
   b. $20 billion
   c. $25 billion
   d. $30 billion

   Ans. c
   p. 303

4. The cost of making cigarettes is _____.
   a. about three cents a pack
   b. about ten cents a pack
   c. about fifty cents a pack
   d. about sixty-three cents a pack

   Ans. a
   p. 303

5. Nicotine kills _____ Americans per year?
   a. 30,000
   b. 300,000
   c. 3,000,000
   d. insufficient evidence

   Ans. b
   p. 306

6. The risk of premature death is _____ higher for smokers than non-smokers.
   a. 20%
   b. 50%
   c. 70%
   d. 90%

   Ans. c
   p. 306

7. A 35-year-old male who smokes two packs a day has a life expectancy that is approximately how many years shorter than a non-smoking counterpart?
   a. 2 years
   b. 4 years
   c. 6 years
   d. 8 years

   Ans. d
   p. 306

8. What is the substance in tobacco that causes dependence?
   a. slovar tar
   b. tar
   c. nicotine
   d. sibray additives

   Ans. c
   p. 306

9. In the 1974 survey, how many workdays are lost in the United States every year by smokers compared to non-smokers?
   a. 12 million
   b. 34 million
   c. 67 million
   d. 81 million

   Ans. d
   p. 307

10. What percentage of lung cancer deaths are caused by smoking? Ans. c
    a. 25% p. 308
    b. 50%
    c. 90%
    d. 98%

11. _____ people in the United States use smokeless tobacco. Ans. c
    a. 2.8 million p. 310
    b. 3.7 million
    c. 5.3 million
    d. 7.1 million

12. Smoke released into the air directly from the lighted tip of a cigarette is called _____. Ans. b
    a. second-hand smoke p. 312
    b. sidestream smoke
    c. tip-top smoke
    d. escaped smoke

13. Recent estimates claim that _____ Americans continue to smoke? Ans. b
    a. 10-20 million p. 315
    b. 50-60 million
    c. 90-100 million
    d. over 100 million

14. What are the two main reasons smokers give for smoking? Ans. c
    a. habit and craving p. 316
    b. manipulation of objects
    c. relaxation and reduction of tension
    d. stimulation and relaxation

15. Which of the following is not a type of behavioral treatment? Ans. d
    a. substitute smoking procedures p. 319
    b. punishment and aversive therapy
    c. stimulus control
    d. impulsive therapy
    e. only a, b, and c

16. Approximately how much does the tobacco industry spend annually on advertising? Ans. d
    a. 25 thousand p. 320
    b. 2.5 million
    c. 25 million
    d. 2.5 billion

17. In China, approximately what percentage of the male population smokes? Ans. d p. 321
    a. 30%
    b. 60%
    c. 75%
    d. 90%

18. Which of the following are considered to be "gateway" drugs? Ans. e p. 322
    a. marijuana
    b. cigarettes
    c. alcohol
    d. cocaine
    e. only a, b, and c

19. _____ percent of people between the ages of 18-25 used some form of tobacco in 1992. Ans. c p. 320
    a. 20 percent
    b. 29 percent
    c. 38 percent
    d. 46 percent

20. As of 1993 _____ of everyone surveyed reported having used cigarettes during their lifetime. Ans. c p. 320
    a. 50 percent
    b. 63 percent
    c. 71 percent
    d. 80 percent

ESSAYS

1. In three paragraphs give a brief history of tobacco use in America. Be sure to include major dates and vital factual information pertaining to tobacco use.

2. Cite then describe four major physiological effects of nicotine.

3. Explain the three types of behavioral treatments used to help people stop smoking. Which do you think is most effective and why?

4. What is smokeless tobacco? Describe second-hand and sidestream smoke.

5. Your best friend argues that smokeless tobacco is much safer to use than smoking; counter argue your friends assertion.

6. What do the most recent health surveys report about tobacco use with regard to gender, education, race and ethnicity?

7. What evidence exists for the finding that daily smokers in comparison to non smokers are much more likely to engage in other drugs? What other drugs in particular?

SUPPLEMENTARY MEDIA

SECOND HAND SMOKE. This sixteen minute video uses a humorous approach to present evidence that secondhand smoke is dangerous to nonsmokers. Follows an evil genius and his unwilling assistant as they discover the dangers of secondhand smoke. Establishes that smoke emitted by a burning cigarette contains more carbon monoxide, nicotine, and tar than smoke inhaled by a smoker. Indicates that secondhand smoke affects the fetus and that children of smokers are sick more often than children of nonsmokers. Intersperses skits throughout the film to illustrate ways nonsmokers can defend their right to breathe clean air. Available from Indiana University Audio-visual Library, Bloomington, Indiana 47405-5901.

SMOKELESS TOBACCO: THE SEAN MARSEE STORY. This sixteen minute video dramatizes the dangers of smokeless tobacco by relating the true story of Sean Marsee, a high school track star and habitual user of snuff, who died of oral cancer at age 19. Interviews Dr. Jim Nethery, a mouth and throat specialist who describes symptoms and stages of oral cancer; Johnnie Johnson, defensive end for the Los Angeles Rams, who talks about professional athletes who use snuff and are trying to quit; and two teenagers, both habitual users of snuff, who discuss its addictive quality and the difficulty they've had quitting. Stresses that smokeless tobacco is not a safe alternative to smoking and shows by example how to say "no" even when faced with peer pressure. Available from Indiana University Audio-visual Library, Bloomington, Indiana 47405-5901.

SMOKING: TIME TO QUIT. This program discusses various ways to stop smoking and the motivation to stop: quitting before or at the onset of pregnancy, when the motivation to protect the unborn child is very strong; a couple quitting together; stop-smoking support groups and their techniques for training ex-smokers to say "No" when a cigarette is offered; being willing to try to quit again after relapsing. (24 minutes, color). Films for the Humanities and Sciences, Princeton, NJ.

CIGARETTES: WHO PROFITS, WHO DIES? For those who believe that the tobacco mortality statistics don't apply to them, here is a message from the folks who make cigarettes that is so grossly cynical it may finally get through. This program features former cigarette models who are now dying of cigarette-related cancer, who were once selected because of their wholesome young looks to persuade others to become addicted to cigarettes. It also shows the new international tactics devised by American tobacco companies in the face of falling demand for their products in this country. (49 minutes, color). Films for the Humanities and Sciences, Princeton, NJ.

SMOKERS ARE HAZARDOUS. The risks of secondary smoke are becoming more and more apparent. This BBC Horizon program documents the facts and the growing international battle. (50 minutes, color). Films for the Humanities and Sciences, Princeton, NJ.

KICK THE HABIT. This program focuses on the effects of cigarette smoking on the body and on the battle against smoking. It shows the efforts being made to educate people to the hazards of smoking, explains the conditioning process by which people become hooked on cigarettes, and presents evidence of the dangers of secondary smoke.(19 minutes, color). Films for the Humanities and Sciences, Princeton, NJ.

– Chapter 12 –

Hallucinogens

The responses to hallucinogens can be highly varied between users and dependent on a person's expectations and surroundings. Generally, these substances alter processing systems of the brain resulting in exaggerated or distorted sensory information. An hallucinogenic experience can be incredibly beautiful and almost spiritual or degrading and terribly frightening. This chapter compares the principal hallucinogenic substances and discusses their history, pattern of use, unique properties and mechanisms of action. Chapter 12 includes sections on (1) the LSD-type hallucinogens such as mescaline, peyote, and psilocybin, (2) amphetamine-type hallucinogens, such as ecstasy and MDA, (3) the anticholinergic hallucinogenic substances such as scopolamine, atropine, the deadly nightshade plant, mandrake and other plants, and (4) the unique hallucinogens such as phencyclidine (PCP), marijuana and inhalants.

TRUE OR FALSE

1. Hallucinogens such as LSD tend to cause physical dependence.
    Ans. F p. 331

2. PCP was originally developed as a general anesthetic.
    Ans. T p. 331

3. Permanent damage to the brain, the heart and other organs can result from using inhalant drugs.
    Ans. T p. 331

4. Hallucinogen abuse has been a major social problem in the U.S. for the past 100 years.
    Ans. F p. 332

5. Timothy Leary was an early opponent of the use of LSD and other hallucinogens.
    Ans. F p. 334

6. Use of hallucinogens is primarily a young adults phenomenon.
    Ans. T p. 334

7. LSD use often causes physical dependence.
    Ans. F p. 335

8. Hallucinogens tend to be less abused than other scheduled drugs because they usually do not cause physical dependence.
    Ans. T p. 335

9. Hallucinogens affect sensory input to the brain.
    Ans. T p. 335

10. Hallucinogens tend to diminish sensory experiences.
    Ans. F p. 336

11. Most often hallucinogen use results in negative experiences. — Ans. F p. 336

12. Hallucinogens can cause effects that are perceived as being an exhilarating spiritual experience. — Ans. T p. 338

13. Some psychiatrists have used hallucinogens on their patients during psychoanalysis. — Ans. T p. 338

14. The psychedelic "peak" of hallucinogens can relate to mystic experiences. — Ans. T p. 338

15. The environment plays a major role in determining a user's response to hallucinogens. — Ans. T p. 338

16. The ergot fungus produces an anticholinergic type of hallucinogen. — Ans. F p. 339

17. LSD almost always causes psychotic behavior that is identical to schizophrenia. — Ans. F p. 340

18. The use of LSD has stayed about the same since the 1960s. — Ans. F p. 341

19. The physical properties of LSD are distinctive. — Ans. F p. 342

20. In its purest form, LSD is colorless, odorless and tasteless. — Ans. T p. 342

21. There are no significant withdrawal symptoms when using LSD. — Ans. T p. 343

22. With the use of LSD, a person can and usually does become physically dependent. — Ans. F p. 343

23. Psychological dependency on LSD can occur. — Ans. T p. 343

24. There is no "typical" pattern of response to LSD. — Ans. T p. 343

25. Tolerance to LSD does not usually occur. — Ans. F p. 343

26. Marijuana use may trigger LSD-related "flashbacks." — Ans. T p. 344

27. Many LSD users find their sense of time distorted. — Ans. T p. 344

28. LSD alters perceptions such that any sensation can be perceived in the extreme. — Ans. T p. 344

29. Flashbacks can usually be controlled by the drug user and only occur when desired.
Ans. F
p. 344

30. LSD has not been proven to cause substantial damage to chromosomes when used by men.
Ans. T
p. 345

31. LSD has not been shown to have carcinogenic or mutagenic effects.
Ans. T
p. 345

32. Most users of peyote prefer natural settings when taking the substance.
Ans. T
p. 347

33. Most mescaline purchased on the streets in big cities is relatively pure.
Ans. F
p. 347

34. Psilocybin was first used by the early natives of Central America more than 2,000 years ago.
Ans. T
p. 347

35. Authentic mescaline is rarely sold in the streets.
Ans. T
p. 347

36. High doses of nutmeg can cause an hallucinogenic effect.
Ans. T
p. 349

37. MDA is sometimes called "the love drug" because it can increase pleasure related to sexual activity and expressions of affection.
Ans. T
p. 350

38. Use of high doses of MDA or ecstacy (MDMA) is likely to cause long-term damage to serotonin neurons in the brain.
Ans. T
p. 351

39. Scopolamine is an anticholinergic drug that can cause hallucinations.
Ans. T
p. 352

40. PCP is currently used on humans as a general anesthetic.
Ans. F
p. 354

41. PCP cannot be easily synthesized in illicit laboratories.
Ans. F
p. 354

42. PCP is currently classified as a Schedule II drug.
Ans. T
p. 354

43. PCP can cause general anesthesia.
Ans. T
p. 354

44. PCP is often sold for other hallucinogenic drugs.
Ans. T
p. 354

45. Intense use of PCP can cause severe psychotic and often violent behavior.
Ans. T
p. 355

46. PCP is sometimes used to treat hospitalized schizophrenic patients.
Ans. F
p. 356

47. Marijuana can aggravate a underlying mental illness such as depression.
Ans. T
p. 357

48. Abusers of nitrous oxide are usually health professionals or their staff.
Ans. T
p. 357

49. The depression caused by high doses of volatile solvents lasts much longer than that caused by alcohol.
Ans. F
p. 357

MULTIPLE CHOICE

1. _____ are substances that cause psychotic-like symptoms.
   a. sympathomimetics
   b. psychotomimetic
   c. psychedelic
   d. ergotism

   Ans. b
   p. 332

2. The League of Spiritual Discovery used _____ as the sacrament.
   a. mescaline
   b. LSD
   c. peyote
   d. none of the above

   Ans. b
   p. 334

3. In 1993, what percentage of high school seniors had ever used hallucinogens?
   a. 2%
   b. 11%
   c. 20%
   d. 30%

   Ans. b
   p. 335

4. In 1975, what percentage of high school seniors had ever used hallucinogens?
   a. 8%
   b. 16%
   c. 24%
   d. 32%

   Ans. b
   p. 335

5. Of the following, which cannot cause hallucinations?
   a. steroids
   b. cocaine
   c. amphetamines
   d. all of these can cause hallucinations

   Ans. d
   p. 335

6. Substances that expand or heighten perception and consciousness are called _____.
   a. photogenetic
   b. psychotomimetic
   c. psychedelic
   d. ergotism

   Ans. c
   p. 335

7. Users of LSD can experience which of these stages?
   a. loss of control
   b. heightened, exaggerated senses
   c. self-reflection
   d. all of the above

   Ans. d
   p. 336-338

8. The recurrence of an earlier drug-induced sensory experience in the absence of the drug is called _____.
   a. "do-over"
   b. "flashback"
   c. "repeat"
   d. "flasher"

   Ans. b
   p. 338

9. Which of the following best describes the synesthesia caused by hallucinogens?
   a. a diminished sense of reality
   b. a sense of being lost and threatened by the experience
   c. another term for "flashbacks"
   d. a cross-over of the senses

   Ans. d
   p. 337

10. Select the incorrect statement concerning the effects of hallucinogens such as LSD.
    a. They can cause vivid and unusual visual and auditory effects.
    b. The user is usually able to control the psychedelic experience.
    c. The user can experience a sense of cosmic merging.
    d. A sense of lost identity caused by using a hallucinogen can be very disturbing to some users.

    Ans. b
    p. 337-338

11. Hallucinogens most likely exert their effects by altering which of the following transmitter systems.
    a. serotonin
    b. dopamine
    c. noradrenalin
    d. epinephrine

    Ans. a
    p. 339

12. Which of the following hallucinogens is not considered to be an LSD-type?
    a. peyote
    b. MDMA
    c. psilocybin
    d. DMT

    Ans. b
    p. 339

13. Select the <u>incorrect</u> statement.
    a. LSD is classified as a phenylethylamine type of hallucinogen.
    b. MDMA has both stimulant and hallucinogenic effects.
    c. High doses of anticholinergic drugs can cause hallucinogenic effects.
    d. Hallucinogens can cause the user to become aware of thoughts and feelings that had been forgotten or repressed.

    Ans. a
    p. 339-340

14. In 1993 what percentage of high school seniors sampled used LSD sometime during their life?
    a. 7%
    b. 13%
    c. 23%
    d. 28%

    Ans. b
    p. 341

15. LSD can be purchased in which of the following forms?
    a. tablets
    b. capsules
    c. liquid
    d. all of these forms can be purchased

    Ans. d
    p. 342

16. When LSD is taken orally, what percentage of the total dose is received by the brain?
    a. 1%
    b. 25%
    c. 75%
    d. 100%

    Ans. a
    p. 342

17. Which of the following is <u>least likely</u> to be caused by LSD?
    a. decreased sympathetic activity
    b. a flood of sensations
    c. increased salivation
    d. increased body temperature

    Ans. a
    p. 342

18. Select the <u>incorrect</u> statement.
    a. LSD is usually taken by mouth.
    b. LSD is sometimes administered by squares of blotter paper.
    c. Large doses of LSD are required to cause hallucinations.
    d. LSD administration activates the sympathetic nervous system.

    Ans. c
    p. 342

19. Which of the following statements concerning LSD's effects is _incorrect_?
    a. Creativity and artistic ability are dramatically increased.
    b. Responses vary considerably for each user.
    c. A person with deep psychological problems is more likely to have an adverse reaction than a person with a well integrated personality.
    d. The psychedelic peak or mystic experience is fast moving and does not occur routinely following LSD use.

    Ans. a
    p. 343-344

20. Which of the following is not considered to be LSD-related flashbacks?
    a. the body trip
    b. the bad mind trip
    c. the sanity trip
    d. altered visual perception

    Ans. c
    p. 344

21. Select the _incorrect_ statement concerning the flashback caused by hallucinogens.
    a. Sensations caused by previous hallucinogen use return.
    b. The experience is always unpleasant and terrifying.
    c. It can consist of altered visual perception including seeing dots, flashes and trails of light.
    d. They can continue to occur for years after hallucinogen use.

    Ans. b
    p. 344

22. Which of the following hallucinogenic substances is found in the peyote cactus?
    a. LSD
    b. DMT
    c. mescaline
    d. psilocybin

    Ans. c
    p. 346

23. _____ is the most active drug in peyote.
    a. LSD
    b. Mescaline
    c. Mydriasis
    d. Methadrine

    Ans. b
    p. 346

24. The definition of mydriasis is _____.
    a. physiological defects
    b. intensified perception
    c. spiritual experiences
    d. pupil dilation

    Ans. d
    p. 347

25. Which of the following is most likely to contain psilocybin?
    a. dried mushrooms
    b. nutmeg
    c. rotted rye
    d. seeds

    Ans. a
    p. 347

26. Select the correct statement concerning the "designer amphetamines."
    a. Use of MDA dulls the sense of touch.
    b. MDA use typically causes LSD-like visual and auditory hallucinations.
    c. Compared to MDA, MDMA has more psychedelic and less stimulant activity.
    d. MDA is also known as "ecstacy."

    Ans. c
    p. 349-350

27. Which of the following spices can be a hallucinogen?
    a. cinnamon
    b. mustard
    c. nutmeg
    d. garlic

    Ans. c
    p. 349

28. Which of the following is referred to as the "love drug."
    a. MDMA
    b. MDA
    c. LSD
    d. Mescaline

    Ans. b
    p. 350

29. Select the incorrect statement concerning anticholinergic hallucinogens.
    a. In ancient societies, these drugs were used as poisons for assassinations.
    b. These drugs are found in many plants that belong to the potato family.
    c. These drugs tend to stimulate salivation.
    d. They work by blocking the activity of the muscarinic receptor.

    Ans. c
    p. 351-352

30. Select the incorrect statement.
    a. Atropine is found in the deadly nightshade plant.
    b. Anticholinergic drugs can cause drowsiness and sedation.
    c. The mandrake plant contains anticholinergic drugs.
    d. Jimson weed is commonly abused by teenagers in this country.

    Ans. d
    p. 352-353

31. Which of the following is least likely to be caused by phencyclidine?
    a. general anesthesia
    b. delirium and psychotic behavior
    c. generalized numbness at low doses
    d. inhibits the sympathetic nervous system

    Ans. d
    p. 354-355

32. Select the incorrect statement concerning PCP.
    a. It is frequently sold as, or substituted for, LSD.
    b. PCP usually causes pleasant and mellow effects.
    c. PCP can give the user a sense of strength, power, and invulnerability.
    d. It can cause perceptual distortions resulting in serious accidents.

    Ans. b
    p. 354-355

33. Drugs with similar structures are called _____.
    a. analogs
    b. replicas
    c. catatonias
    d. volatiles

    Ans. a
    p. 355

34. Which of the following statements concerning PCP toxicity is incorrect?
    a. Diagnosis of a PCP toxicity is frequently missed because the symptoms closely. resemble an acute schizophrenic episode
    b. There are specific antagonists available to treat acute overdoses of PCP.
    c. Valium is often used to sedate the agitation caused by PCP.
    d. Long-term use can cause vague cravings after cessation of PCP.

    Ans. b
    p. 356-357

35. A state of oxygen deficiency is called _____.
    a. volatile
    b. hypoxia
    c. catatonia
    d. analog

    Ans. b
    p. 357

36. Select the incorrect statement concerning the inhalant chemicals of abuse
    a. The inhalant chemicals of abuse can cause hallucinations and euphoria.
    b. They include some solvents, glue, hairspray, lighter fluid and gasoline.
    c. These chemicals are classified as Schedule II substances by the DEA.
    d. Chronic inhalation can cause damage to heart, brain and kidneys.

    Ans. c
    p. 357-359

37. Which of the following is <u>not likely</u> to be caused by chronic use of the volatile solvents? Ans. c p. 359
    a. memory impairment
    b. motor impairment
    c. increased appetite
    d. severe depression

ESSAYS

1. Name the four stages of the LSD experience.

2. List four reasons that the use of LSD has diminished since the 1960s.

3. Identify the three principal types of hallucinogens and how they differ.

4. Give a one or two paragraph outline on the history of hallucinogen use. Be certain to include dates and names of important people and events in your answer.

5. Why were substances with hallucinogenic properties so popular with ancient religions and cults? Give examples.

6. Why have hallucinogens been so interesting to psychiatrists and psychologists?

SUPPLEMENTARY MEDIA

ACID. This film discusses what is known about LSD. Available from Kent State Film Library, Kent, Ohio 43242.

FOCUS ON DRUGS: LSD AND OTHER PSYCHEDELICS. This fifteen minute film covers the effects of LSD and similar hallucinogens on the body and mind. Available from Boston University, Krasker Memorial Film Library, 565 Commonwealth Ave., Boston, MA 02215

ANGEL DEATH. This film discusses the dangerous drug PCP. The drug causes a great deal of damage and is often disguised as some other type of drug. It often results in serious violent behavior and in paranoid behavior. Available from Indiana University, Audio-visual Library, Bloomington, Indiana 47505-5901.

— Chapter 13 —

MARIJUANA

The chapter begins with some recent attitudes regarding marijuana use followed by an interesting analysis of the history of marijuana use. In the next two sections the characteristics of marijuana use are discussed followed by a detailed section on the physiological effects of marijuana use. This section includes: effects on the central nervous system; respiratory system; cardiovascular system; sexual performance and reproduction; immune system; tolerance and dependence, and therapeutic uses.

Related to the physiological effects of marijuana use is a discussion of the behavioral effects of marijuana use. This section discusses driving performance, chronic use, the amotivational syndrome, as well as subjective experiences. The final section discusses some interesting trends in marijuana use; such as age differences, peer influences, other research findings and the role of marijuana as a gateway drug.

TRUE OR FALSE

1. In 1993 there was a reversal in the declining use of marijuana. — Ans. T p. 366

2. Marijuana is one of the largest cash-producing crops. — Ans. T p. 367

3. Highly regarded governmental officials and judges have recommended that use of marijuana be legalized. — Ans. T p. 367

4. Marijuana plants have been used to make rope and clothing. — Ans. T p. 367

5. It has been proven that use of marijuana can make people commit murder. — Ans. F p. 369

6. Marijuana is currently classified as a narcotic. — Ans. F p. 369

7. During World War II the Federal Government subsidized the growth of marijuana. — Ans. T p. 369

8. Cannabis sativa is the hemp plant marijuana. — Ans. T p. 370

9. The lack of pollination of the female marijuana plant by the male plant dramatically increases the potency of the marijuana. — Ans. T p. 371

10. There are no cancer-causing ingredients in marijuana. — Ans. F p. 372, 373

11. When marijuana is smoked, it takes several minutes for the active ingredient to reach the brain. Ans. F p. 372

12. Marijuana can cause stimulation, depression, and hallucinations. Ans. T p. 372-373

13. Most marijuana smokers retain more tar residue in their lungs than tobacco smokers. Ans. T p. 373

14. Chronic smoking of marijuana can cause severe pulmonary damage. Ans. T p. 373

15. Cannabis has several effects on semen. Ans. T p. 374

16. Marijuana use by pregnant women impairs fetal growth. Ans. T p. 374

17. Tolerance rarely occurs to the effects of marijuana. Ans. F p. 375

18. Severe addiction to marijuana is relatively rare. Ans. T p. 376

19. Marijuana has been prescribed by physicians in order to combat nausea associated with chemotherapy. Ans. T p. 376

20. THC has been approved by the FDA to stimulate appetite in AIDS patients. Ans. T p. 376

21. Physicians have prescribed marijuana to be used as a muscle relaxant and a way to remedy depression. Ans. T p. 377

22. There is no evidence that use of marijuana impairs a person's ability to drive. Ans. F p. 378

23. Cannabis appears to be a _relatively_ safe substance. Ans. T p. 380

24. Amotivational syndrome characterizes regular users of marijuana who experience an increase in motivation and productivity. Ans. F p. 380

25. The extent of Marijuana use is strongly related to age. Ans. T p. 382

26. Metropolitan area residents use marijuana more than nonmetropolitan residents. Ans. T p. 383

27. Those who have completed less years of education are more likely to use marijuana casually.   Ans. F p. 383

28. The unemployed are more likely to use marijuana than those employed.   Ans. T p. 383

29. At low doses, marijuana has a sedative effect.   Ans. T p. 386

30. It is unlikely that a person will use marijuana if his/her peers don't.   Ans. T p. 383

31. Females are more likely to use marijuana intensely that their male counterparts.   Ans. F p. 383

32. Since 1979, marijuana use has increased for all ages.   Ans. F p. 386

33. Much evidence points to marijuana being the number one "gateway" drug in America.   Ans. F p. 386

34. The effectiveness of marijuana decreases with repeated use.   Ans. T p. 386

MULTIPLE CHOICE

1. What percentage of high school seniors have used marijuana?   Ans. b p. 366
   a. 12%
   b. 35%
   c. 62%
   d. 82%

2. How tall was the world's largest marijuana plant?   Ans. d p. 370
   a. 9 1/2 feet
   b. 17 feet
   c. 28 feet
   d. 39 feet

3. _____ is a type of marijuana plant that means "without seeds."   Ans. b p. 371
   a. Cannabis sativa
   b. Sinsemilla
   c. Acapulco Gold
   d. Marinol

4. The weakest form of marijuana is _____.   Ans. a p. 372
   a. bhang
   b. ganja
   c. hashish
   d. cannabis

5. Smoking a marijuana cigarette can alter _____.
    a. mood
    b. coordination
    c. memory
    d. all of the above

    Ans. d
    p. 372

6. Hunger experienced while under the effects of marijuana is referred to as (the) _____.
    a. "cravies"
    b. "eaties"
    c. "munchies"
    d. "pangos"

    Ans. c
    p. 372

7. The main psychoactive chemical in marijuana is:
    a. glaucoma
    b. ergot
    c. THC
    d. none of the above

    Ans. c
    p. 372

8. How long does it take to eliminate the psychoactive chemical in marijuana from body fat?
    a. up to 24 hrs.
    b. up to 5 days
    c. up to 20 days
    d. up to 30 days

    Ans. d
    p. 372

9. The strongest derivative of marijuana is _____.
    a. ganja
    b. sinsemilla
    c. hashish
    d. bhang

    Ans. c
    p. 372

10. How long does it take for the main psychoactive chemical in marijuana to reach the brain after inhalation?
    a. 3-5 seconds
    b. 14-15 seconds
    c. 30-35 seconds
    d. over 1 minute

    Ans. b
    p. 372

11. A substance that stimulates or intensifies sexual desire is called _____.
    a. altered perceptions
    b. angina pectoris
    c. mycardia
    d. aphrodisiac

    Ans. d
    p. 374

12. FDA approved THC in capsule form is referred to as _____.
    a. Cannabis
    b. Sinsemilla
    c. Pectoris
    d. Marinol

    Ans. d
    p. 376

13. _____ is an eye disease manifested by increased intraocular pressure, and _____ relieves this problem.
    a. Pinkey; magic mushrooms
    b. Glaucoma; marijuana
    c. Polar Vision; marijuana
    d. Angina; amphetamines

    Ans. b
    p. 376

14. Medically, THC has been used as a (an)_____.
    a. muscle-relaxant
    b. antiseizure compound
    c. antidepressant
    d. all of the above

    Ans. d
    p. 376-377

15. In 1989, of the 1,800 tested, what percentage of drivers arrested for driving intoxicated, tested positive for marijuana?
    a. 2%
    b. 19%
    c. 29%
    d. 42%

    Ans. b
    p. 378

16. To date, how many people have been killed by an overdose of marijuana (alone)?
    a. none have been reported
    b. less than 500
    c. around 1000
    d. over 1000

    Ans. a
    p. 380

17. Who is more likely to use marijuana?
    a. 12-16 year olds
    b. 18-26 year olds
    c. 35 and older
    d. all ages equally

    Ans. b
    p. 382

18. Select the incorrect statement concerning marijuana use:
    a. Approximately equal numbers of males and females have used marijuana.
    b. Residents of large metropolitan areas are more likely to use marijuana than persons in small towns.
    c. There is no correlation between the rate of marijuana use and perceived harmfulness.
    d. Access to marijuana does not appear to be as important in determining use as peer pressure.

    Ans. c
    p. 383

19. Who is more likely to use marijuana intensely?
    a. males
    b. females
    c. upper class
    d. Protestants

    Ans. a
    p. 383

20. Which group of people were <u>least likely</u> to   Ans. b
    have used marijuana?   p. 383
    a. Northerners
    b. Southerners
    c. Easterners
    d. Westerners

21. Which group of people were <u>most likely</u> to have   Ans. d
    used marijuana?   p. 383
    a. Northerners
    b. Southerners
    c. Easterners
    d. Westerners

ESSAYS

1. Define in three or more sentences the following: *hashish, ganja, sensimilla,* and *bhang.*

2. Has the potency of marijuana changed in the last ten years? If so, how and why?

3. Why are many governmental and law enforcement officials in favor of legalizing marijuana?

4. What particular effects does marijuana have on the central nervous system, respiratory system and the cardiovascular system?

5. How has marijuana been used therapeutically?

6. What medicinal evidence has been discovered regarding chronic use of marijuana?

7. Cite five trends you found particularly interesting regarding marijuana use?

SUPPLEMENTARY MEDIA

MARIJUANA AND HUMAN PHYSIOLOGY. This twenty-one minute video explores the damaging effects marijuana has on the body and dispels the belief that smoking pot is just harmless recreation. Describes how marijuana's chief psychoactive ingredient, tetrahydrocannabinol (THC), damages the sinuses, pharynx, uvula, lungs, heart, brain, reproductive system, immune system, and cell division. Talks with former marijuana users who discuss the physiological, psychological, and emotional consequences of using the drug. Looks at the development of the "pot personality," the dangers of driving while under the influence, and the hazards of mixing marijuana with alcohol. Available from the Indiana University Audio-visual Library, Bloomington, Indiana 47405-5901.

MARIJUANA, DRIVING AND YOU. This thirteen minute 16mm film compares the effects of a marijuana high with the perceptions and judgements needed for safe, defensive driving. Relates how the feeling of euphoria, narrowed field of awareness, and altered sense of time deny drivers full command of their driving abilities. Considers also the danger of drugs being added to the marijuana, the unpredictability of marijuana's effects, and the risks of mixing marijuana with drugs of any kind. Available from Indiana University Audio-visual Library, Bloomington, Indiana 47405-5901.

– Chapter 14 –

Drugs and Therapy

Prescription and nonprescription (OTC) drugs have been viewed differently by the public since these classifications were established in 1951. However, distinctions between these drug categories have become somewhat blurred by recent changes in public demand and federal policies. An example of the somewhat arbitrary nature of prescription and nonprescription classifications is the policy of the Food and Drug Administration to switch effective and relatively safe prescription medication to OTC status. In this Chapter we discuss and compare these two drug groups. We begin by examining the policies of OTC drug regulation followed by a discussion of safe self-care with non prescription drug products. A short explanation of some of the most popular medications in this category conclude the section on OTC drugs. The second part of this chapter is a general overview of prescription drugs. The consequence of misuse of prescription drugs as well as ways for patients to avoid such problems are discussed. A brief presentation of some of the most commonly prescribed drugs conclude this chapter.

TRUE OR FALSE

1. The OTC drug market is projected to reach almost $30 billion by the year 2010.
   Ans. T
   p. 392

2. About 75% of Americans routinely self-medicate with OTC medications.
   Ans. T
   p. 392

3. Approximately $1 billion is spent annually on OTC medications in the U.S.
   Ans. F
   p. 392

4. Approximately 300,000 OTC drug products are available OTC in the U.S.
   Ans. T
   p. 392

5. Most principal ingredients still included in OTC drug products are category III.
   Ans. F
   p. 393

6. Cough suppressants are the most commonly used OTC medications.
   Ans. F
   p. 394

7. Several major health problems have resulted from switching prescription drugs to OTC status.
   Ans. F
   p. 394

8. Ibuprofen is a drug that was switched from prescription to OTC status.
   Ans. T
   p. 394

9. It is not likely that the FDA will switch any more drugs from prescription to OTC status in the future.
   Ans. F
   p. 394

10. Labels of OTC drug products must indicate if the active ingredients are classified as category I or III.
Ans. F
p. 396

11. Excessive use of some OTC drugs can cause dependence.
Ans. T
p. 396

12. Pharmacists are a good source for information concerning OTC drug products.
Ans. T
p. 398

13. OTC analgesics are most effective in the treatment of pain associated with internal organs.
Ans. F
p. 399

14. It requires a lower dose of Ibuprofen to relieve arthritis than to relieve a muscle ache.
Ans. F
p. 399

15. Acetaminophen has antipyretic effects.
Ans. T
p. 399

16. Acetaminophen should not be used by children because it causes Reye's syndrome.
Ans. F
p. 399

17. Excessive use of some OTC drugs can cause structural damage to the body.
Ans. T
p. 400

18. Caffeine is able to relieve some types of pain.
Ans. T
p. 400

19. Some OTC analgesics contain the caffeine equivalent of one-fourth to one-half cup of coffee per tablet.
Ans. T
p. 400

20. Children of 1 to 5 years of age are the most susceptible to colds.
Ans. T
p. 400

21. Sympathomimetics are frequently used as decongestants in OTC cold medications.
Ans. T
p. 402

22. Excessive use of topical decongestants can cause congestion rebound.
Ans. T
p. 402

23. Use of large quantities of Vitamin C has been proven to prevent the common cold.
Ans. F
p. 403

24. OTC sleep aids are often very effective in the treatment of chronic insomnia.
Ans. F
p. 404

25. "Look alike" stimulants can cause headaches, breathing problems and rapid heartbeat when used in high doses.
Ans. T
p. 404

26. "Look alike" stimulants are sometimes used as substitutes for amphetamines.
Ans. T
p. 404

27. Antacids are the most commonly used OTC medications for gastrointestinal problems. — Ans. T p. 405

28. Excessive use of calcium carbonate can cause rebound acidity in the stomach if its use is stopped abruptly. — Ans. T p. 406

29. Usually excessive gastric acidity can be successfully treated with OTC antacids. — Ans. T p. 405

30. Magnesium salts are used as an antacid in low doses and a laxative in high doses. — Ans. T p. 406

31. Antihistamines are the active anorexiants used in OTC diet aids. — Ans. F p. 407

32. About 1 million cases of skin cancer in the U.S. are caused by exposure to U.V. rays. — Ans. T p. 409

33. Exposure to the sun's U.V. rays helps to diminish wrinkles in the skin. — Ans. F p. 409

34. People with a fair skin complexion should use sunscreens with lower SPF numbers for adequate skin protection. — Ans. F p. 410

35. Dentists are allowed to write prescriptions for drugs. — Ans. T p. 411

36. A generic drug is usually not as effective as its proprietary counterpart. — Ans. F p. 412

37. Narcotic analgesics are often combined with aspirin or acetaminophen for better pain-relieving actions. — Ans. T p. 415

38. The prescription NSAIDS are like Ibuprofen in their pharmacological action. — Ans. T p. 415

39. Antibiotics effective against many species of bacteria are designed as narrow spectrum drugs. — Ans. F p. 415

40. Antibiotics that kill bacteria are designed as bacteriostatic. — Ans. F p. 415

41. Prozac has been found to be much more dangerous for most depressed patients than other antidepressant medication. — Ans. F p. 416

42. Approximately 1% of the population in the U.S. has some form of epilepsy. — Ans. T p. 417

43. Cardiovascular disease is the number one leading cause of death in the U.S. — Ans. T p. 419

44. Almost 1/3 of the top selling 30 prescription drugs are for the treatment of cardiovascular diseases.
    Ans. T
    p. 419

45. Angina pectoris is caused by an excessive flow of blood to the heart.
    Ans. F
    p. 420

46. High doses of niacin are sometimes used to lower cholesterol.
    Ans. T
    p. 420

MULTIPLE CHOICE

1. Approximately how much money is spent on OTC medications in the U.S.?
   a. $2 billion
   b. $6 billion
   c. $12 billion
   d. $20 billion

   Ans. c
   p. 392

2. Approximately how many OTC drugs products are available in the U.S.?
   a. 50,000
   b. 100,000
   c. 300,000
   d. 1,000,000

   Ans. c
   p. 392

3. Most ingredients still in OTC drug products have been classified in which of the following categories?
   a. I
   b. II
   c. III
   d. IV

   Ans. a
   p. 393

4. Which of the following is the most commonly used OTC medication in the U.S.?
   a. cold medicines
   b. antihistamines
   c. antacids
   d. pain relievers (analgesics)

   Ans. d
   p. 394

5. Of the current top 10 OTC medications how many had been switched from prescription?
   a. none
   b. 2
   c. 5
   d. 9

   Ans. d
   p. 394

6. Which of the following is not required by the FDA to be included on OTC drug labels?
   a. approved uses
   b. classification of active ingredients (i.e., category I, II, or III)
   c. instructions for safe and effective use
   d. cautions and warnings

   Ans. b
   p. 396

7. Which of the following OTC analgesics has the least anti-inflammatory effects?
   a. aspirin
   b. acetaminophen
   c. salicylate
   d. ibuprofen

   Ans. b
   p. 399

8. Which of the following pains is <u>least likely</u> to be relieved by OTC analgesics?
   a. headache
   b. muscle ache
   c. tooth ache
   d. stomach cramps

   Ans. d
   p. 399

9. Which of the following drugs is classified as a nonsteroidal antiinflammatory (NSAID)?
   a. salicylate
   b. codeine
   c. antihistamines
   d. decongestants

   Ans. a
   p. 399

10. Which of the following best describes "antipyretics?"
    a. drugs that relieve inflammation
    b. drugs that relieve pain
    c. drugs that cause nausea
    d. drugs that reduce fevers

    Ans. d
    p. 399

11. Which of the following is a side effect of acetaminophen?
    a. causes Reye's syndrome in children
    b. irritates the stomach
    c. interferes with blood clotting
    d. none of the above

    Ans. d
    p. 400

12. Approximately how much caffeine (in terms of quantity in a cup of coffee) is included in a tablet of OTC combination analgesics such as Anacin and Excedrin?
    a. one-tenth
    b. one-fourth to one-half
    c. one
    d. two

    Ans. b
    p. 400

13. The common cold accounts for what percentage of the acute illnesses in the U.S. ?
    a. 1%
    b. 5%
    c. 20%
    d. 40%

    Ans. c
    p. 400

14. Common OTC decongestant cold products contain which of the following ingredients?
    a. sympathomimetics
    b. analgesics
    c. antihistamines
    d. all of the above

    Ans. d
    p. 400-401

15. How do sympathomimetics relieve nasal congestion associated with colds and allergies?
    a. they cause vasoconstriction
    b. they cause vasodilation
    c. they stimulate deeper breathing
    d. they relax muscles in the air passages

    Ans. a
    p. 402

16. Select the incorrect statement:
    a. antitussives should be used to treat productive coughs
    b. codeine is classified as an antitussive
    c. expectorants stimulate mucus production and decrease its viscosity
    d. hard candy often can reduce minor coughing

    Ans. a
    p. 402

17. Which of the following drugs are commonly used in OTC sleep aids?
    a. antitussives
    b. antihistamines
    c. sympathomimetics
    d. caffeine

    Ans. b
    p. 404

18. Which of the following drugs is not likely to be included in "look-alike" stimulants?
    a. caffeine
    b. phenylpropanolamine
    c. ephedrine
    d. codeine

    Ans. d
    p. 404

19. Which of the following is the most frequently OTC medication used to treat gastrointestinal disorders?
    a. laxatives
    b. antacids
    c. drugs to relieve diarrhea
    d. anti-nausea medication

    Ans. b
    p. 405

20. Which of the following OTC antacids is most likely to cause formation of carbon dioxide gas in the stomach?
    a. sodium bicarbonate
    b. calcium carbonate
    c. aluminum hydroxide
    d. magnesium salts

    Ans. a
    p. 405

21. Which of the following are used as "anorexiants" in OTC diet aids? Ans. c
    a. antihistamines p. 407
    b. analgesics
    c. sympathomimetics
    d. antitussives

22. OTC drugs that cause the keratin layer of the skin to peel away are used for which of the following purposes? Ans. b
    a. increase the softness of the skin p. 409
    b. to treat acne
    c. remove skin blemishes
    d. remove body hair

23. Which of the following medical problems are caused by exposure to UV rays? Ans. d
    a. premature aging in the skin p. 409
    b. deadly melanomas
    c. squamous cell carcinomas
    d. all of the above

24. What does the "SPF" number on OTC sunscreen products stand for? Ans. d
    a. it is an indication of the toxicity of the product p. 410
    b. it is the FDA's stamp of approval for the product
    c. it indicates the amount of protection against UV-A rays
    d. it indicates the amount of protection against UV-B rays

25. Which of the following is not a criterion for a drug to be restricted by prescription? Ans. a
    a. intended to treat self-limiting diseases p. 410
    b. can cause addiction
    c. to be used for diseases requiring the supervision of health professional
    d. new and without an established safe track-record

26. Which of the following health professionals do not have any prescription writing privileges? Ans. c
    a. dentists p. 411
    b. physicians assistants
    c. chiropractors
    d. pharmacists

27. Which of the following best describes the legislative act called OBRA '90?
    a. laws regulating the quality of prescription drugs
    b. laws giving pharmacists the right to prescribe drugs
    c. laws requiring pharmacists to give patients necessary information on proper prescription drug use
    d. laws restricting the access by pharmacists to prescription drugs that are especially addicting

    Ans. c
    p. 412

28. Why are generic drugs usually less expensive than their proprietary counterpart?
    a. they are a lesser quality
    b. the generic company did not invest in the discovery and development of the drug
    c. the generic companies are usually more efficiently operated
    d. generic drugs usually contain smaller quantities of the expensive ingredients

    Ans. b
    p. 412

29. Antibiotics are most effective in the treatment of infections caused by which of the following microorganisms?
    a. viruses
    b. fungi
    c. protozoa
    d. bacteria

    Ans. d
    p. 415

30. Severe depression afflicts what percent of the population at any one time?
    a. 1-2%
    b. 5-6%
    c. 10-12%
    d. 15-20%

    Ans. b
    p. 415

31. What is the likely cause of most severe cases of depression?
    a. neurotransmitter imbalances in the brain
    b. malnutrition
    c. no known underlying cause
    d. personality defect

    Ans. a
    p. 416

32. Select the _incorrect_ statement concerning diabetes:
    a. Type I diabetes only occurs during childhood
    b. Type I diabetics are treated with insulin injections
    c. Insulin medication is classified according to its duration of action
    d. Type II diabetics are often treated with oral hypoglycemic

    Ans. a
    p. 417

33. Which of the following best describes the antiulcer effects of drugs like Tagamet and Zantac?   Ans. d
    p. 419
    a. They neutralize gastric acidity
    b. They cause relaxation and relieve stress
    c. They have antibiotic activity
    d. They block gastric secretions

34. Bronchodialators are used for which of the following therapeutic purposes?   Ans. d
    p. 419
    a. relieve congestion associated with the common cold
    b. increases heart efficiency
    c. increase urine formation
    d. relieve asthma attacks

35. Which of the following drugs are most likely to be used to treat hypertension?   Ans. a
    p. 419
    a. diuretics and vasodialators
    b. cardiostimulators and vasoconstrictors
    c. bronchodilators and antiarrhythmics
    d. bronchoconstrictors and antianginals

ESSAYS

1. Discuss the potential problems of making more effective OTC drugs available to the public.

2. What are the general differences between prescription and non prescription drugs?

3. Cite three of the five FDA guidelines for antacid use.

4. Define the following in two or more sentences and give examples where appropriate: antitussives; expectorants; electrolytes; anorexiants; and keratin layer.

5. Describe the advantages and disadvantages of switching drugs from prescription to non prescription status.

6. Discuss the rules for safe and effective use of OTC medication.

7. Discuss the type of questions a patient should ask when given a drug prescription.

SUPPLEMENTARY MEDIA

ADDICTION CAUSED BY MIXING DRUGS (GX 3179-VHS). This 19 minute film video discusses addiction caused by mixing nonaddicting prescription drugs. Available from "Films for the Humanities and Sciences." P.O. Box 2053, Princeton, NJ 08543 or call (800)257-5126.

DRUGS AND MEDICINES: WHAT IS MISUSE? This ten minute 16mm film uses a pharmacy setting to dramatize the common ways drugs and medicines are misused. Points out the dangers of taking someone else's prescription, increasing the dosage, or extending the duration of the prescribed treatment. Suggests that most people are not aware that they are abusing drugs. Available from Indiana University Audio-visual library, Bloomington, Indiana 47405-5901.

OVER-THE-COUNTER DRUGS: SMOOTH TALK AND SMALL PRINT. This film introduces the viewer to consumerism for OTC drugs and discusses the way to make intelligent selections for these drugs. Available from the University of Wisconsin, Lacrosse, Film Rental Library, 1705 State Street, Lacrosse, Wisconsin 54601.

OVER-THE-COUNTER PILLS AND PROMISES. This film is a continuation of background on selection of OTC drugs. Comparisons are important in selecting the best drug for the least money. Available from the University of Nevada, Reno, film library. Getchell Library, Reno, Nevada 89557 is thirty minute 16mm film examines prescription drug abuse through interviews with doctors and patients, concentrating on the creation of "legal" addicts by physicians who are quick to prescribe tranquilizers for symptoms of stress. Looks at the expensive promotional techniques used by drug companies to encourage doctors to prescribe mood altering drugs for a variety of non-existent diseases, such as the empty nest syndrome, car-itis, and picky eater-itis. Interviews doctors who state that physicians need to take more time with their patients and use community service groups to help counsel patients as alternatives to tranquilizers. Includes interviews with a current tranquilizer user, a former user, and a user who had been simultaneously addicted to drugs and alcohol. Available from Indiana University Audio-visual Library, Bloomington, Indiana 47405-5901.

Chapter 15

Drug Abuse Among Special Populations

There is no such thing as a typical "drug abuser." Drug abuse problems are biased by the environmental-emotional configuration of the user. To effectively identify the nature of substance dependence and its consequences it is essential to understand the psychosocial, and biological makeup, of those who become addicted to these drugs. Chapter 15 examines four selective populations who, because of their unique features, present special drug abuse problems in the terms of identification, prevention, and treatment. Adolescents are discussed relative to their unique developmental status. In particular, the psychological, social and environmental patterns that encourage substance abuse as well as adolescent suicide, violence and gang involvement are examined. Drug abuse and women comprise the next section of this chapter. The unique feature of substance abuse and its impact on women are discussed. Problems with substance abuse during pregnancy are also explained. The third population presented in this chapter are athletes. Why, what and how drugs are abused by athletes are the main themes of this section. The final discussion of Chapter 15 relates to drug abuse and AIDS. I.V. drug users have the second highest rate of HIV infection in this country. The reasons for the high evidence of HIV in drug abusers, the consequences and preventive measures are explained.

TRUE OR FALSE

1. Any use of substances of abuse by adolescents should be aggressively treated.
   Ans. F
   p. 428

2. Emotionally stable adolescents who relate well to peers and family are less likely to have drug abuse problems than those with emotional disturbances and low self-esteem.
   Ans. T
   p. 428

3. Most adolescent use of substances of abuse will not progress to major abuse problems.
   Ans. T
   p. 429

4. The home is not a major factor in determining whether an adolescent will abuse drugs.
   Ans. F
   p. 429

5. Adolescent drug use increased from 1980 to early 1990's.
   Ans. F
   p. 430

6. Adolescent drug abuse patterns are very different from those of adults.
   Ans. T
   p. 431

7. Experts believe that drug abuse by adolescents is more likely to be a symptom, rather than a cause, of social maladjustment.
   Ans. T
   p. 431

8. Adolescents who abuse alcohol and other drugs often feel insecure and inferior and have a self-destructive attitude.
    Ans. T
    p. 432

9. Adolescent alcoholics have a suicide rate more than 50 times greater than the national average.
    Ans. T
    p. 432

10. Most adolescents experiment with drugs due to antisocial and deviant behavior.
    Ans. F
    p. 432

11. Often, the psychosocial deficiencies that encourage adolescents to become involved in gangs are the same factors that lead to drug abuse.
    Ans. T
    p. 433-434

12. Many of the behavioral patterns that occur coincidentally with drug problems also occur in adolescents who are not involved in drugs.
    Ans. T
    p. 436

13. If therapy for drug abuse is to be successful the environment of the adolescent usually must be improved.
    Ans. T
    p. 438

14. Most drug-abuse programs have facilities especially designed to care for adolescent patients.
    Ans. F
    p. 439

15. Research on drug abuse is usually conducted equally in male and female populations.
    Ans. F
    p. 439-440

16. In 1992 female college students were more likely than their male counterparts to use cigarettes daily.
    Ans. T
    p. 440

17. Use of cocaine during pregnancy increases the likelihood of miscarriage.
    Ans. T
    p. 442

18. Women are more likely than men to develop severe alcohol dependence.
    Ans. F
    p. 443

19. Women are likely to initiate their drinking patterns earlier in life than men.
    Ans. F
    p. 443

20. Health consequences for comparably excessive alcohol use is more severe for men than women.
    Ans. F
    p. 444

21. Female alcoholics are more likely than male alcoholics to have alcoholic parents or siblings.
    Ans. F
    p. 444

22. Women are more likely than men to abuse prescription drugs because they are more likely to suffer mental disorders such as depression and anxiety.
Ans. T
p. 444

23. Athletes are more likely to use marijuana and alcohol than nonathletes.
Ans. F
p. 446

24. Approximately 2% of college males use anabolic steroids.
Ans. T
p. 448

25. Women athletes do not use anabolic steroids.
Ans. F
p. 448

26. The likelihood of using anabolic steroids by professional athletes is less than by college athletes.
Ans. F
p. 449

27. The use of different steroids taken singly but in sequence is called "cycling."
Ans. T
p. 449

28. Power athletes (football players) prefer "cycling" steroids, while body builders prefer "stacking."
Ans. F
p. 450

29. Increases in muscle mass disappear once the use of anabolic steroids stops.
Ans. T
p. 451

30. About 50% of anabolic steroids used in the U.S. are obtained from the black market.
Ans. T
p. 452

31. Stimulants are most likely to improve performance if the athlete is fatigued.
Ans. T
p. 452-453

32. Most competitions do not allow the use of caffeine by the athletes.
Ans. F
p. 453

33. Some competitions do not allow the use of OTC stimulants such as those found in decongestant drugs.
Ans. T
p. 453

34. Erythropoietin is easily detectable in the urine of athletes who use this drug.
Ans. F
p. 453

35. More than 50% of all AIDS patients are I.V. drug users.
Ans. F
p. 455

36. Testing for HIV infection is reliable within 2-5 days of the infection.
Ans. F
p. 456

37. If conducted properly, testing for HIV infections is very reliable.
Ans. T
p. 456

38. HIV cannot be passed prenatally from an infected mother to a child.
Ans. F
p. 457

39. AIDS is the leading killer of black males from 25 to 44 years of age.
Ans. T
p. 457

MULTIPLE CHOICE

1. Select the <u>incorrect</u> statement:
   a. Alcoholic parents frequently have parents who abuse drugs
   b. A parent who abstains from drinking and has a rigid moralistic approach to life almost never has drug abusing children
   c. Parents who have unrealistic expectations of their children are more likely to raise children at high risk for substance abuse
   d. Overly protective parents often create an environment that encourages substance abuse by their children

   Ans. b
   p. 429-430

2. Which of the following sociological factors is <u>least likely</u> to encourage adolescent drug use?
   a. rejection from family members
   b. rejection from peers
   c. minority race in a well integrated tolerant society
   d. membership in a gang that views drug use as desirable

   Ans. c
   p. 430

3. The rate of first drug exposure of U.S. 8th graders in 1992 was approximately:
   a. 5-10%
   b. 15-20%
   c. 30-40%
   d. 50-60%

   Ans. b
   p. 430

4. Select the statement that most accurately describes the adolescent drug abuse pattern in 1993.
   a. declining dramatically compared to 1992 levels
   b. disturbing increases in drug abuse compared to 1992 especially with marijuana
   c. no changes observed compared to 1992
   d. no surveys conducted in 1993, so no patterns determined

   Ans. b
   p. 430

5. Select the incorrect statement concerning ethnic differences in 1992 drug use patterns among adolescents.
   a. Eighth grade hispanics tended to have the highest prevalence of drug use.
   b. Whites in the eighth grade were less likely than blacks to abuse drugs.
   c. Of seniors in high school, hispanics were most likely to use the most dangerous drugs, such as cocaine, crack and heroin.
   d. Smoking rates for blacks were about 20% of those for whites.

   Ans. b
   p. 430

6. Select the incorrect statement concerning adolescent drug abuse patterns:
   a. Adolescents who abuse drugs are more likely to engage in criminal activity than adults who abuse drugs.
   b. It is more likely that other members of the family abuse drugs when the adults are abusers rather than adolescents.
   c. Adolescent, more than adult, drug abusers are likely to be associated with a dysfunctional family that engages in emotional or physical abuse of its members.
   d. Adolescents, more than adults, are likely to begin their drug use because of peer pressure.

   Ans. b
   p. 431

7. Which of the following populations in the U.S. is most likely to commit suicide?
   a. black males 14 to 20 years
   b. white females 14 to 20 years
   c. hispanic males 19 to 24 years
   d. white males 14 to 20 years

   Ans. d
   p. 432

8. Select the incorrect statement:
   a. Alcohol is the most significant drug factor associated with date, acquaintance and gang rapes by adolescents
   b. 40-50% of the female adolescents who abuse drugs have been the victims of sexual abuse
   c. Almost half of the offenders consume alcohol before molesting a child
   d. Only 20-30% of child molesters abused drugs as adolescents

   Ans. d
   p. 432-433

9. Select the incorrect statement concerning gangs: Ans. b
   a. Adolescents in a gang usually have similar socioeconomical, racial and ethnic backgrounds
   b. Gang membership is loosely defined in terms of appearance and dress
   c. Leadership and seniority of a gang are usually defined by tenure, age and achievements
   d. A strong family environment and guidance from parents or guardians are deterrents to gang involvement

   p. 433-434

10. Select the incorrect statement concerning treatment of drug abuse in adolescents: Ans. b
    a. Drug abuse is less likely if adolescents are involved in productive, structured groups or clubs (e.g., athletic or fine arts groups and teams)
    b. Any use of substances of abuse should be viewed as abnormal behavior and not tolerated
    c. Adolescents should be taught that drugs are never the solution for emotional difficulties nor are they useful long-term coping techniques
    d. Adolescents are less likely to abuse drugs if they understand that these substances deprive them of independence and control over their lives

    p. 436

11. Select the incorrect statement concerning U.S. female drug abuse patterns in 1992: Ans. b
    a. Female college students were less likely to use heroin than males
    b. Annual prevalence rates (use at least one time during the year) for using stimulants of abuse were much lower for female than male college students
    c. Daily use of alcohol by male college students was almost double that of their female counterparts
    d. Lifetime rates of marijuana use were higher in white than black females

    p. 440

12. Which of the following most accurately describes the relative occurrence of lung cancer in male and female cigarette smokers? Ans. c
    a. more likely to occur in males than females
    b. incidence is not different in males and females
    c. more likely to occur in females than males
    d. not known because the study has not been conducted

    p. 442

13. Select the <u>incorrect</u> statement concerning drug use during pregnancy:
    a. Use of cocaine in late stages of pregnancy can cause a stroke and brain damage to the fetus.
    b. Use of high doses of alcohol during pregnancy may cause facial damage to the fetus.
    c. The incidence of fetal alcohol syndrome in the U.S. is 3-5% of newborns.
    d. Consumption of high doses of caffeine during pregnancy can interfere with fetal development.

    Ans. c
    p. 442-443

14. In 1992, what percent of American women, ages 19-32 years, had used alcohol sometime during their life?
    a. 80-90%
    b. 50-60%
    c. 30-40%
    d. 15-25%

    Ans. a
    p. 443

15. Which of the following is <u>most likely</u> the explanation for the observation that greater adverse effects tend to occur in female than male alcoholics?
    a. female alcoholics consume more alcoholic beverages
    b. female alcoholics tend to consume drinks with higher alcoholic content
    c. female alcoholics have slower alcohol metabolism and smaller blood volume
    d. the female organs are more sensitive than the corresponding male organs and more likely to be damaged by alcohol

    Ans. c
    p. 444

16. Which of the following accounts for the reluctance of women to seek treatment for drug dependence?
    a. they have domestic roles with high expectations
    b. treatment centers usually do not meet their unique female health requirements
    c. they are often unemployed and can't afford private, personalized care.
    d. All of the above

    Ans. d
    p. 445

17. Which of the following drugs are athletes more likely to abuse than nonathletes?
    a. alcohol
    b. cocaine
    c. barbiturates
    d. hallucinogens
    e. amphetamines
    f. all of the above

    Ans. e
    p. 446

18. Which of the following is classified as anabolic steroids? Ans. b
    a. estrogens p. 447-448
    b. androgens
    c. adrenalin
    d. dopamine

19. Approximately how many Americans have used or are currently using anabolic steroids to improve athletic performance? Ans. c
    a. less than 100,000 p. 448
    b. 500,000
    c. 1 million
    d. 10 million
    e. 20 million

20. Anabolic steroids have been classified in which of the following Schedules? Ans. c
    a. I p. 448
    b. II
    c. III
    d. IV
    e. V

21. Which of the following best describes "stacking" of anabolic steroids by athletes? Ans. d
    a. The use of several steroids taken singly but in sequence p. 449
    b. The use of several steroids with an overlaying dosing pattern
    c. The alternating use of steroids and amphetamines
    d. The use of several types of steroids at the same time

22. Which of the following is least likely to be an effect of anabolic steroids? Ans. a
    a. They greatly enhance the strength and athletic skills of the average adult p. 451
    b. Increased risk of liver disease
    c. Increased incidences of irritability and outburst of anger
    d. Breast enlargement in males and breast reduction in females
    e. Infertility in both sexes

23. Select the *incorrect* statement:
    a. Erythropoietin is used as a substitute for "blood doping"
    b. Athletes use human growth factor to try and stimulate muscle growth
    c. Beta-adrenergic blockers are used by archers to reduce the tremors associated with nervousness
    d. Gamma-hydroxybutyrate(GHE) usually causes drowsiness

    Ans. d
    p. 453-454

24. Select the *incorrect* statement concerning AIDS:
    a. The AIDS condition often occurs within a few weeks after HIV infection
    b. HIV damages the immune system by destroying $CD4+$-type helper T lymphocytes
    c. A brief flu-like illness usually occurs 6 to 12 weeks after exposure to HIV
    d. Even though asymptomatic, the HIV-infection person is still contagious

    Ans. a
    p. 455-456

25. Which of the following is *not* a route of transmission of HIV?
    a. blood transfusion
    b. use of contaminated needles
    c. mosquitos
    d. semen and vaginal fluid

    Ans. c
    p. 457

26. Select the *incorrect* statement concerning AIDS:
    a. More men than women die from AIDS
    b. Almost 3/4 of the female AIDS victims are infected with HIV because of behavior related to drug abuse
    c. Because crack is consumed by inhalation and not I.V. injection, its users have a much lower rate of HIV infection than I.V. cocaine users
    d. There is a high incidence of HIV infection in cocaine "shooting galleries"

    Ans. c
    p. 457

ESSAYS

1. Discuss the unique psychological status of adolescents that makes them particularly vulnerable to drug abuse problems.

2. What types of parents are most likely to raise children at high risk for substance abuse?

3. What types of adolescents are most likely to attempt suicide?

4. Describe the behaviors that can be warning signals that drug abuse problems are present in an adolescent.

5. Explain why women are less likely than men to seek professional help for their drug abuse problems.

6. Describe how motherhood confounds treatment of women for drug abuse problems.

7. Why does a societal double standard exist for alcoholism in women and men?

8. Explain why many athletes are willing to take the risks associated with drug abuse.

9. Why is there a high rate of HIV infection in the drug abusing population?

10. Describe approaches to prevent the spread of AIDS.

SUPPLEMENTARY MEDIA

DOWNFALL: SPORTS AND DRUGS (1988). Audience: teenagers, length: 30 minutes. This documentary videotape profiles athletes whose careers were destroyed by drugs. NCADI (P.O. Box 2345, Rockville MD 30847-2345; VHS 13.

PRIVATE VICTORIES (1988)- Audience: teenagers. Length 116 minutes - four 29 minute episodes. These videotapes emphasize that young people are better off without drugs. The episodes show how deciding against drugs can influence friends to do the same. NCADI (P.O. Box 2345, Rockville MD 20847-2347: VHS 15).

AMERICA IN JEOPARDY: THE YOUNG EMPLOYEE AND DRUGS IN THE WORKPLACE (1992) - Audience: employees and employers. Length: 20 minutes. This video contains interviews of recovering drug users and warns that taking drugs is a dead end, and mixing drugs with work is a big mistake. The host explains how drugs affect the body and the mind. NCADI (P.O. Box 2345, VHS 44).

ADOLESCENT TREATMENT APPROACHES (1992) - Audience: Adults. Length: 25 minutes. This video discusses the problems of adolescents entering drug treatment and the techniques that are most likely to be successful. The video stresses the importance of understanding the special needs of adolescents for successful treatment. It also evaluates the role of the family in successful treatment of drug dependence. NCADI (P.O. Box 2345, Rockville MD 30847-2347; VHS 40).

TREATMENT ISSUES FOR WOMEN (1992) - Audience: Adult. Length: 22 minutes. This video presents the specific issues, methods and techniques used to treat drug-dependent women. It discusses the unique challenges of treating women who use drugs and presents ways to deal with relationship building, sexual and physical abuse, anger and the role of confusion. NCADI (P.O. Box 2345, Rockville MD 30847-2345: VHS 39).

– Chapter 16 –
Drug Education, Prevention, and Treatment

This chapter tackles the vast array of educational, preventive, and treatment-oriented drug use and/or abuse programs, procedures, and modalities that need to address divergent drug users. Major view, controversies, strengths and weaknesses of prevention and treatment is also discussed.

After briefly discussing how pervasive the use of drugs is and some reasons for its use, the first major section discusses the nature of drug addiction, views on, and levels of addiction. The next section embarks on drug education and how it can be used to control addiction, create program models, and how curriculum-based drug education can produce achievable objectives. The next major section delves into drug prevention. Three approaches to prevention programs targeted to children and adolescents are discussed, along with the role of social and personal development, treating alternatives to drugs, and some important nontraditional prevention methods are included.

Drug treatment comprises the last section of this chapter. Treatment is discussed according to three types of settings, namely, in hospital programs, residential care, and day hospitals and intensive outpatient services. Next, are major treatment approaches: (1) medical treatments -- detoxification and abstinence programs, maintenance programs, morphine maintenance, methadone maintenance and opiate antagonists; (2) psychological and psychotherapeudic approaches; (3) social approaches -- alcoholics anonymous (AA), therapeutic communities, and family therapy; (4) behavioral approaches -- operant or instrumental conditioning, contingency management techniques, counterconditioning (sensitization), biofeedback relaxation therapy, and alternative behavioral approaches; and (5) innovative treatments -- hypnosis and acupuncture. Finally the last part of this concluding section discusses evaluating program effectiveness.

TRUE OR FALSE

1. Successful rehabilitation has to be specific to the type of person and the type of drug user.　　Ans. T p. 465

2. Addiction is only one phase of drug use.　　Ans. T p. 465

3. Drug education actually began in the early 1900s with prohibition.　　Ans. F p. 465

4. Maintenance programs eliminate drug addiction.   Ans. F p. 465

5. Opiates and cocaine were illegal in the United States during the 1800s.   Ans. F p. 467

6. Opiates and cocaine were illegal in the United States since the 1800s.   Ans. F p. 467

7. Habitual use is the third stage of the addiction process.   Ans. F p. 468

8. The belief that people abuse alcohol or drugs because they choose to do so is part of the Moral Model.   Ans. T p. 468

9. A person who frequently uses chemicals during or for recreation is at Level 0 of drug addiction.   Ans. F p. 471

10. Most drug use begins as a result of social influence.   Ans. T p. 472

11. For committed users, drug education should aim to prevent or delay drug abuse.   Ans. T p. 473

12. The cognitive model of education stresses accurate and straight information on drug abuse.   Ans. T p. 474

13. Primary prevention entails using immediate intervention once drug use has begun.   Ans. F p. 475

14. Social inoculation theory states that youth need to be "armed" against the alcohol and drug attack.   Ans. T p. 476

15. NIDA stands for National Intravenous Drug Abusers.   Ans. F p. 477

16. Exercise, diet, and scuba diving are all plausible alternatives for a drug abuser.   Ans. T p. 480

17. Modifying the body's chemistry is a way to alter conscious.   Ans. T p. 482

18. Medically, drug abuse is considered an illness.   Ans. T p. 484

19. Acute care refers to serious or critical care.   Ans. T p. 485

20. Methadone is a noneuphoric opiate.   Ans. T p. 486

21. The concept of maintenance on a noneuphoric opiate is not very acceptable in the United States. — Ans. F p. 486

22. Maintenance programs provide a steady amount of a noneuphoric drug to the user. — Ans. T p. 487

23. Colonidine suppresses the symptoms of hallucinogenic withdrawal. — Ans. F p. 487

24. An antagonist is a compound that enhances the actions of a drug. — Ans. F p. 487

25. The psychological model views drug use as a cultural phenomenon. — Ans. F p. 488

26. Alcoholics Anonymous is a national organization emphasizing responsible drinking. — Ans. F p. 489

27. Therapeutic communities are programs that advocate a complete change in lifestyle. — Ans. T p. 491

28. A token economy system is a system of rewards and punishments used to manipulate or condition a behavior and responses. — Ans. T p. 492

29. Reinforcemtent is any behavior that strengthens the likelihood that a behavior will not be repeated. — Ans. F p. 492

30. Alternative approaches consist of substitutions chemically induced states of euphoria with natural "highs." — Ans. T p. 493

31. Biofeedback relaxation therapy is also referred to as sensitization. — Ans. F p. 493

32. Natural states of consciousness are stressed in the alternative behavior approaches. — Ans. T p. 493

33. The arousal state of consciousness is called ergotropic. — Ans. T p. 495

34. Less sensory input enters the brain during the trophotropic stage than during the ergotropic stage. — Ans. T p. 495

35. A conscious state similar to hallucination is called catatonic. — Ans. F p. 495

36. A person can be taught to hypnotize himself/herself to gain further control when s/he is tempted to use drugs. — Ans. T p. 496

37. Hypnosis has been used to link drug-taking behavior to negative consequeces such as anxiety. Ans. T p. 496

38. The success of a drug treatment program can be based on abstinence from illegal drugs, the patient's ability to hold down a job, and his/her improved physical and mental health. Ans. T p. 497

MULTIPLE CHOICE

1. Which is <u>not</u> a primary approach for treating drug abuse? Ans. d p. 465
   a. medical
   b. innovative
   c. social
   d. educational

2. It is estimated that well over _____ of all Americans have used some kind of drug. Ans. b p. 466
   a. 20%
   b. 30%
   c. 40%
   d. 50%

3. Researchers at the University of Michigan have concluded that general drug use has _____ since the 1800s. Ans. a p. 466
   a. declined
   b. remained about the same
   c. increased
   d. changed drastically from year-to-year

4. Over _____ million Americans use drugs monthly. Ans. a p. 466
   a. 23
   b. 35
   c. 47
   d. 83

5. In the United States, drug dependence has been seen as a(n) _____. Ans. d p. 467
   a. criminal act
   b. moral violation
   c. illness
   d. all of the above
   e. only a and c

6. The four phases of the addiction process are: Ans. d p. 467
   a. initial, habitual, addiction and remorse
   b. casual, habitual, social and addiction
   c. beginning, intermediate, obligatory, final
   d. habitual, initial addiction and relapse

7. Which is not a characteristic of addiction? Ans. b
   a. chronic use p. 468
   b. deviant behavior
   c. compulsion
   d. resulting problems

8. _____ of drug law violators released from prison Ans. b
   were rearrested for a second offense within a p. 470
   year.
   a. 10 percent
   b. 25 percent
   c. 35 percent
   d. 50 percent

9. Drug dependence is often associated with _____. Ans. d
   a. divorce p. 470
   b. domestic violence
   c. AIDS
   d. all of the above

10. Medical complications such as ulcers, fatty Ans. c
    deposits, and hepatitis usually begin to occur p. 471
    during _____ of drug addiction.
    a. Level 1
    b. Level 2
    c. Level 3
    d. Level 4

11. How many levels of addiction can be identified? Ans. c
    a. 4; levels 0-3 p. 471
    b. 4; levels 1-4
    c. 5; levels 0-4
    d. 5; levels 1-5

12. Using persuasion to dissuade someone from using a Ans. a
    particular drug is an example of _____. p. 472
    a. primary prevention
    b. secondary prevention
    c. tertiary prevention
    d. none of the above

13. Drug education began in the _____. Ans. a
    a. 1800s p. 472
    b. 1920s
    c. 1950s
    d. 1980s

14. Formal education about drug use began in the _____.   Ans. d
    a. 1920s   p. 472
    b. 1940s
    c. 1960s
    d. none of the above

15. The _____ model stresses accurate and unbiased   Ans. b
    information.   p. 474
    a. information-based
    b. cognitive
    c. peer tutoring
    d. social resistance training

16. Recreational drug use, legal versus illegal   Ans. c
    drugs, and positive and negative role models are   p. 475
    usually discussed in drug education programs at
    the _____.
    a. elementary level
    b. junior high level
    c. senior high and college level
    d. none of the above

17. Which is not a type of prevention?   Ans. a
    a. auxillary   p. 475
    b. tertiary
    c. secondary
    d. primary

18. Prevention used to help individuals stop abusing   Ans. c
    drugs immediately is _____ prevention.   p. 476
    a. primary
    b. secondary
    c. tertiary
    d. none of the above

19. DARE was developed by _____.   Ans. a
    a. Los Angelas Police Department   p. 476
    b. National Coalition for Drug Abuse
    c. Special Office for Drug Abuse Prevention
    d. Nancy Reagan

20. Since 1989, the federal government has spent _____   Ans. c
    dollars on funding new drug education programs.   p. 477
    a. 350 million
    b. 750 million
    c. 3 billion
    d. 5 billion

21. The exploration of positive alternatives to drug abuse is called the _____ approach.
    a. cognitive
    b. reinforcement
    c. alternative
    d. repitition
    Ans. c
    p. 480

22. Solitary confinement, biofeedback, and meditation are examples of _____.
    a. reducing sensory input
    b. increasing sensory input
    c. seclusion/inclusion therapy
    d. none of the above
    Ans. a
    p. 482

23. The first drug treatment program, SAODAP, was established in _____.
    a. 1963
    b. 1966
    c. 1969
    d. 1971
    Ans. d
    p. 484

24. The first drug treatment program, the Special Action Office for Drug Abuse Prevention (SAODAP) was established in _____.
    a. 1963
    b. 1970
    c. 1975
    d. none of the above
    Ans. d
    p. 484

25. Active treatments for drug addiction have been around for _____ years.
    a. 10
    b. 20
    c. 30
    d. 40
    Ans. b
    p. 484

26. Detoxification is usually a _____.
    a. short-term program
    b. long-term program
    c. short-term or long-term program, depending on the level of addiction
    d. none of the above
    Ans. a
    p. 485

27. Which is not a major treatment approach?
    a. social
    b. psychological
    c. innovative
    d. isolative
    Ans. d
    p. 485

28. Maintenance programs are examples of _____ drug treatment method.
    a. psychological
    b. psycho therapeutic
    c. social behavioral
    d. medical

    Ans. d
    p. 486

29. Detoxification programs have a goal of _____.
    a. stabilizing drug addiction
    b. reducing drug addiction to zero
    c. determining the level of addiction
    d. none of the above

    Ans. b
    p. 486

30. As of 1987, _____ heroin addicts were on methadone maintenance.
    a. 70,000
    b. 90,000
    c. 100,000
    d. 120,000

    Ans. a
    p. 486

31. Which is not an opiate antagonist?
    a. cyclazocine
    b. nalorphine
    c. methadone
    d. naloxone

    Ans. c
    p. 487

32. When was Alcohol Anonymous (AA) founded?
    a. mid 1920s
    b. mid 1930s
    c. mid 1940s
    d. mid 1950s

    Ans. b
    p. 489

33. Therapeutic communities are an example of _____.
    a. an in hospital program
    b. residential care
    c. an intensive outpatient service
    d. none of the above

    Ans. b
    p. 491

34. Programs that advocate a complete change in lifestyle are called _____.
    a. therapeutic communities
    b. self-regulating communities
    c. both a and b
    d. none of the above

    Ans. c
    p. 491

35. In 1990, how many residential therapeutic communities serving drug abusers were there?
    a. 100
    b. 200
    c. 300
    d. 400

    Ans. d
    p. 491

36. Alcohologists are _____. Ans. d
    a. people in AA programs p. 491
    b. people who teach in AA programs
    c. recovered alcoholics
    d. people who research alcohol addiction

37. As of 1990 there were over _____ residential TC's. Ans. c
    a. 100 p. 491
    b. 200
    c. 400
    d. 800

38. Behavior that strengthens the likelihood that a behavior will be repeated is called _____. Ans. c
    a. stimulus p. 492
    b. alternative approach
    c. reinforcement
    d. repetition

39. Which is not a behavioral approach? Ans. b
    a. operant conditioning p. 492
    b. restructuring therapy
    c. biofeedback therapy
    d. contingency management techniques

40. Token economy systems have been shown to greatly _____ attendance or participation at drug therapy sessions. Ans. b
    a. decrease p. 493
    b. increase
    c. stabilize
    d. it has shown no results at all

41. The highest level of stimuli coming into the brain during an ergotropic state is when the nervous system is _____. Ans. c
    a. aroused p. 494
    b. hyperaroused
    c. ecstatic
    d. none of the above

42. Catatonia may occur when the nervous system is: Ans. b
    a. aroused p. 494
    b. hyperaroused
    c. ecstatic
    d. none of the above

43. What state of conscious is achieved when there is complete union with the absolute or the self? Ans. d
    a. catotonic p. 495
    b. mystical rapture
    c. satori
    d. yoga samadhi

44. Oxygen consumption can be decreased as much as _____ in a person who meditates. Ans. b
    a. 10% p. 495
    b. 20%
    c. 30%
    d. 40%

45. Which is **not** a basic condition required to elicit a relaxation response? Ans. d
    a. a quiet environment p. 496
    b. an object to dwell on
    c. a comfortable position
    d. a strong tranquilizing drug

46. Acupuncture has been practiced in _____ for at least 3000 years for medical purposes. Ans. a
    a. China p. 496
    b. Japan
    c. Australia
    d. England

47. Most detoxification programs last no longer than _____ days. Ans. c
    a. 7 p. 498
    b. 14
    c. 21
    d. 28

ESSAYS

1. What are the five levels of drug addiction?

2. List and briefly discuss the three different types of program models for drug education programs.

3. List and define the five major treatment approaches.

4. Give three reasons why you think Alcoholics Anonymous has been so successful in treating alcoholics.

5. List and discuss the four basic conditions to elicit relaxation.

SUPPLEMENTARY MEDIA

OUR WONDERFUL BODY: MEDICINES, DRUGS AND POISONS. This film introduces the proper use of medicines, drugs, and poisons, including the importance of reading labels and following the doctor's recommendations. Explains that overtiredness and not eating properly can often cause illness and emphasizes that the proper use of medicine can help to cure illnesses. Discusses the safe use of common poisons and household chemicals and urges that all medicines, drugs, and poisons be stored in a safe area. Available from Indiana University Audio-visual Library, Bloomington, Indiana 47405-5901.

WHOSE BODY IS IT ANYWAY? This film examines the effects of advertising, for both prescription drugs and over-the-counter products, in creating the self-perpetuating belief that there must be a chemical answer for every problem. Notes the negative psychological attitude found in medical patients, who feel cheated if they leave without a prescription, and indicates that one quarter of the drugs prescribed today are tranquilizers. Discusses the importance of comparing the cost and quality of different drugs, the risks versus the benefits of using drugs, and the patient's responsibilities and rights. Available from Indiana University Audio-visual Library, Bloomington, Indiana 47405-5901.

RELAPSE PREVENTION (1992). Audience: adult. Length: 24 minutes. What do we really know about relapse, and how can we help clients avoid it? This videotape addresses these questions by providing information on the phenomenon of relapse and its often chronic appearnace in the lives of alcohol and other drug abusers. The accompanying user's guide gives information on the components of relapse prevention. (VHS37) NCADI Catolog fall/winter 93/94 Rockville, MD 20847.

TREATMENT ISSUES FOR WOMEN (1992). Audience: adult. Length: 22 minutes. Why is it important to identify, understand, and treat the special needs of drug-dependent women? Specific issues, methods, and techniques are presented to help viewers understand the multifaceted dimensions of treating women who use drugs. This videotape examines the challenges that women bring to treatment, along with ways to deal with relationship building, sexual and physical abuse, anger, and role confusion. Scenes from several innovative treatment programs show how treatment services have been enhanced to serve women. The users' guide lists resources for clinicians. (VHS39) NCADI Catalog fall/winter 93/94 Rockville, MD 20847.

ADOLESCENT TREATMENT APPROACHES (1992). Audience: adult. Length: 25 minutes. What kinds of problems do adolescents bring with them when they enter treatment, and what techniques will best treat them? The videotape stresses the importance of understanding the specific needs that accompany adolescents' development as the key to success in treatment. This understanding begins with accurate assessment and continues with aftercare monitoring. The videotape examines the family's role in treating alcohol and other drug problems among adolescents. Family therapy and other treatment progrmas are also presented. Clinicians are encouraged to broaden their perspectives in particular topical areas through the user's guide. (VHS40) NCADI Catalog fall/winter 93/94 Rockville, MD 20847.